→INTRODUCING

INFINITY

BRIAN CLEGG & OLIVER PUGH

ICON

This edition published in
the UK and the USA
in 2012 by Icon Books Ltd,
Omnibus Business Centre,
39–41 North Road, London N7 9DP
email: info@iconbooks.net
www.introducingbooks.com

Sold in the UK, Europe and Asia
by Faber & Faber Ltd,
Bloomsbury House,
74–77 Great Russell Street,
London WC1B 3DA or their agents

Distributed in South Africa
by Book Promotions,
Office B4, The District,
41 Sir Lowry Road,
Woodstock 7925

Distributed in Australia and
New Zealand by
Allen & Unwin Pty Ltd,
PO Box 8500,
83 Alexander Street,
Crows Nest, NSW 2065

Distributed to the trade in the USA
by Consortium Book Sales
and Distribution
The Keg House,
34 Thirteenth Avenue NE, Suite 101,
Minneapolis, MN 55413-1007

Distributed in Canada
by Penguin Books Canada,
90 Eglinton Avenue East,
Suite 700, Toronto,
Ontario M4P 2Y3

ISBN: 978-184831-406-1

Originating editor: Richard Appignanesi

Printed and bound in Great Britain by Clays Ltd, St Ives plc

Big numbers

Infinity, as no end of people will tell you, is a big subject. It will take you into history, philosophy and the physical world, but is best first approached through mathematics. It makes sense to ease into it via **big numbers**.

By giving a lengthy number a name you seem to demonstrate your power over it – and the bigger the number is, the more impressive your ability. This is reflected in the reported early life of Gautama Buddha. As part of his testing as a young man in an attempt to win the hand of Gopa, Gautama was required to name numbers up to a huge, totally worthless value. Not only did he succeed, but he carried on to bigger numbers still.

100,000,000,000,000,000? EASY, THAT'S ACHOBYA.

Googoled

It's fine giving names to numbers we encounter every day, but how many of us will ever use *this number*?

10,000,000,000,000,000,000,000,000,000,000,000,00
0,000,000,000,000,000,000,000,000,000,000,000,
000,000,000,000,000,000,000,000,000,000,000

As it happens, it does have a name, one that proved a problem for the unfortunate Major Charles Ingram when it was his million-pound question on TV show *Who Wants to be a Millionaire?* He was asked if the number – 1 with 100 noughts after it – was a "googol", a "megatron", a "gigabit" or a "nanomol". Major Ingram favoured the last of these, until a cough from the audience prompted him towards googol. To be honest, who can blame him? "Googol" sounds childish.

Googol *is* childish – for a good reason. In 1938, according to legend, mathematician Ed Kasner was working on some numbers on his blackboard at home. His nephew, nine-year-old Milton Sirrota, was visiting. Young Milton spotted the biggest number and is supposed to have said: "That looks like a googol!"

This isn't a very convincing story, though. There's no reason why Kasner would bother to write such a number on a blackboard.

WHAT WOULD YOU CALL A REALLY, *REALLY* BIG NUMBER (SAY 1 WITH 100 NOUGHTS AFTER IT)?

A GOOGOL!

Symbols from India

To deal with any number we need **symbols** that represent numerical values. The symbol equivalents of the words "one", "two", "three" and so on (1, 2, 3…) arrived in the West from India via the Arabic world. The oldest known ancestors of the modern system were found in caves and on coins around Bombay dating back to the 1st century AD.

The numbers 1 to 3 were based on a line, two lines and three lines, like horizontal Roman numerals, though they can still be seen with some imagination in the main strokes of our modern numbers. The markings for 4 to 9 are closer ancestors of the symbols we use today.

The Indian symbols were adopted in the Arabic world, coming to the West in the 13th century thanks to two books, written by a philosopher in Baghdad and a traveller from Pisa. The earlier book, lost in the Arabic original, was written by **al-Khwarizmi** (c. 780–850) in the 9th century. The Latin translation of this, *Algoritmi de numero Indorum*, was produced around 300 years later, and is thought to have been considerably modified in the process.

The version of al-Khwarizmi's name in the title is usually given as the origin of the term "algorithm", though it's sometimes linked to the Greek word for number, *arithmos*.

The Book of Calculation

The traveller from Pisa was **Leonardo Fibonacci** (c. 1170–1250). (His father, a Pisan diplomat, was Guglielmo Bonacci, and "Fibonacci" is a contraction of *filius Bonacci*, son of Bonacci.) He travelled widely in North Africa and became the foremost mathematician of his time, his name inevitably linked to the Fibonacci numbers (see page 15), which he popularized but didn't discover. Although *Numero Indorum* was translated into Latin a little before Fibonacci's book *Liber abaci* came out in 1202, it seems that *Liber abaci* ("The Book of Calculation") had the bigger influence in introducing the Indian system to the West.

ON MY TRAVELS I WAS INTRODUCED TO THE ART OF THE INDIAN'S NINE SYMBOLS.

0, a powerful tool

The symbols we use for numbers are arbitrary. ¶, ß, √, π, л would do as well as 1, 2, 3, 4, 5. However, the new Indian numerals brought with them a very powerful tool. Earlier systems from Babylonian through to Roman were **tallies**, sequential marks to count objects. We're most familiar with Roman numerals – the tally sequence is obvious in I, II, III, IV, V – where V is effectively a crossed through set of IIII and IV is one less than V. But the trouble with such systems is that there's no obvious mechanism to add, say, XIV to XXI.

THE NEW SYSTEM USED COLUMNS WITH A PLACE-HOLDER *O* FOR EMPTY SPACES, TRANSFORMING ARITHMETIC.

Archimedes: *The Sand Reckoner*

But whatever symbols are used, big numbers kept their appeal. In a book called *The Sand Reckoner*, ancient Greek philosopher **Archimedes** (c. 287–212 BC) demonstrated to King Gelon of Syracuse that he could estimate the number of grains of sand it would take to fill the universe.

We don't know a lot about Archimedes, but we do have a number of his books, which show him to be a superb mathematician and a practical engineer. He is said to have devised defence weapons for Syracuse ranging from ship-grabbing cranes to vast metal mirrors to focus sunlight and set ships on fire.

Unlike many of Archimedes' other works, *The Sand Reckoner* wasn't exactly practical. But there was a serious point behind this entertaining exercise. What Archimedes set out to do was to show how the Greek number system, which ran out at a myriad myriads (100 million), could be extended without limit. He first estimated the size of the universe at around 1,800 million kilometres (just outside the orbit of Saturn).

"UNIVERSE" IS THE NAME GIVEN BY MOST ASTRONOMERS TO THE SPHERE WHOSE CENTRE IS THE CENTRE OF THE EARTH.

He then decided how many sand grains are needed to be the size of a poppy seed, how many of these fit in a sphere of finger's breadth, and so on up to fill the universe, using his newly designed system. His final count suggested that the universe should hold around 10^{51} sand grains (1 with 51 zeros after it).

Tantalizingly, Archimedes also mentions the work of the philosopher Aristarchus, who had written a book (now lost) that put the Sun at the centre of the universe rather than the Earth. Archimedes calculated the size of this universe too, which he made significantly bigger than the traditional model. Aristarchus' fanciful Sun-centred universe would hold around 10^{63} grains.

THERE ARE SOME, KING GELON, WHO THINK THAT THE NUMBER OF THE SAND IS INFINITE IN MULTITUDE ...

The poetry of infinity

Archimedes' feat was later celebrated by the poet **John Donne** (1572–1631), who commented: "Men have calculated how many particular graines of sand, would fill up all the vast space between the Earth and the Firmament." Donne used this huge number to emphasize that it was negligible in contrast with the limitless nature of infinity and eternity.

BUT IF EVERY GRAIN OF SAND WERE THAT NUMBER, AND MULTIPLIED AGAIN BY THAT NUMBER, YET ALL THAT MADE UP NOT ONE MINUTE OF THIS ETERNITY ...

The Sand Reckoner could also have been the inspiration for the opening of the poem "Auguries of Innocence" by **William Blake** (1757–1827): "To see a World in a Grain of Sand / And a Heaven in a Wild Flower, / Hold Infinity in the palm of your hand / And Eternity in an hour."

Number sequences

In practice, Archimedes had used only a tiny fraction of his system, but the Greeks were also aware of **sequences** of numbers that went on for ever. Number sequences are part of every culture. Most children recite counting rhymes ("One, two, buckle my shoe …") to help recall the sequence of the counting numbers.

Once children learn the basic structure of the numbers and the way the sequence of integers* works, they often count up and up interminably. But where do the numbers stop? Children often seem to be trying to find the biggest number. But they'll never get there. They could count for the rest of their life, and there would still be as many numbers to go as there were to start with. Imagine there were a biggest number, let's call it *max*. What's to stop us adding *max*+1, *max*+2 and so on? The dance never ends.

305, 306, 307 ... 549, 550, 55

.. 810, 811, 812 ... max, max+

max+2, max+3, max+4, max+

max+6, max+7, max+8, max+

 *Words marked with an asterisk are explained in the Glossary on page 172.

Of course the counting numbers aren't the only simple number sequence that most of us would recognize. You can make a sequence by doubling the previous number:

$$1, 2, 4, 8, 16, 32, 64\ldots$$

Or you can have sequences with a back-and-forth alternation of steps, for example:

$$1, 3, 2, 4, 3, 5, 4, 6, 5, 7\ldots$$

There's the Fibonacci sequence, and others relying on adding previous numbers:

$$1, 2, 3, 5, 8, 13, 21\ldots$$

Or sequences where multiplication is involved:

$$1, 2, 2, 4, 8, 32, 256, 8192\ldots$$

And there's no need to stick to whole numbers. As far back as the ancient Greeks there has been an awareness of sequences of fractions, such as:

$$1, \frac{1}{2}, \frac{1}{3}, \frac{1}{4}, \frac{1}{5}, \frac{1}{6}\ldots$$

Strange sequences

At first sight, chains that go on for ever seem harmless, but it doesn't take long to find some that are strange. In a series* we add the numbers up as we go along to produce a sum. Take a very simple series, alternating 1 and –1:

$$1 - 1 + 1 - 1 + 1 - 1 \ldots$$

It's hardly rocket science. Each 1 is cancelled out by a –1, so the total of the series is 0:

$$(1 - 1) + (1 - 1) + (1 - 1) \ldots = 0$$

Or is it? Just shift the brackets and we still have a series that cancels out, but now we've got a 1 left over:

$$1(-1 + 1)(-1 + 1)(-1 + 1) \ldots = 1$$

So the same series adds up to both 0 *and* 1. This has been rephrased as: "If you turn a light bulb on and off an infinite set of times, does it end up on or off?" It could be either. This is a mathematician's answer – a physicist will tell you that it's off, because the bulb has blown.

Or take another simple series where each item is half the last:

$$1 + \tfrac{1}{2} + \tfrac{1}{4} + \tfrac{1}{8} = 1\tfrac{7}{8}$$

It seems, as we add in element after element …

$$1 + \tfrac{1}{2} + \tfrac{1}{4} + \tfrac{1}{8} + \tfrac{1}{16} = 1\tfrac{15}{16}$$

… that it's going to eventually reach 2:

$$1 + \tfrac{1}{2} + \tfrac{1}{4} + \tfrac{1}{8} + \tfrac{1}{16} + \tfrac{1}{32} = 1\tfrac{31}{32}$$

… though in practice with any particular number there's always a little gap left:

$$1 + \tfrac{1}{2} + \tfrac{1}{4} + \tfrac{1}{8} + \tfrac{1}{16} + \tfrac{1}{32} \ldots + \tfrac{1}{MAX} = 1\tfrac{(MAX-1)}{MAX}$$

You could say that the series adds up to 2 if you have an infinite set of components – but what does that mean? And how can an infinite number of things add up to a finite quantity?

The infinity machine

In 1949, the German mathematician and physicist **Hermann Weyl** (1885–1955), a contemporary of Einstein, devised an imaginary "infinity machine", inspired by this sequence. Such a machine would carry out a sequence of steps, taking (say) 1 second for the first step, ½ second for the second step, ¼ second for the third step and so on. In principle it could undertake an infinite sequence of steps in a finite time.

There seem to be two difficulties in practice, though.

... AND THE OTHER IS WHETHER TIME CAN TRULY BE SPLIT INTO INFINITELY SMALL SEGMENTS.

ONE IS MAKING ANYTHING PHYSICAL HAPPEN IN AN INCREASINGLY SHORT TIME ...

Zeno's paradoxes

This series $1 + \frac{1}{2} + \frac{1}{4} + \frac{1}{8} + \frac{1}{16} \ldots$ was the basis of one of Zeno's famous paradoxes. Greek philosopher **Zeno of Elea** (c. 490–430 BC) belonged to the school of Parmenides, which considered reality to be unchanging and movement to be an illusion. Zeno knocked up a number of entertaining examples to demonstrate the faulty nature of our attitude to change and motion. Probably the best known is the arrow that encourages us to imagine two arrows. One floats stationary in space. The other is flying at full speed. Now catch them at a snapshot in time.

HOW DO WE TELL THE DIFFERENCE?

HOW DOES ONE ARROW KNOW TO MOVE IN THE NEXT FRACTION OF TIME WHILE THE OTHER DOESN'T?

Achilles and the tortoise

But the paradox that reflects our sequence of 1 + ½ etc. concerns Achilles and the tortoise. This unlikely pair are setting
out on a race. Achilles, being after all a hero, gives the slower
tortoise a lead. And they're off. In a very small amount of time,
Achilles has reached the tortoise's position. But by then, the tortoise has moved on. In an even shorter amount of time, Achilles
has reached the tortoise's new spot. And again it has moved on.

IT DOESN'T MATTER HOW
MANY TIMES YOU GO
THROUGH THIS PROCEDURE –
AN INFINITE SET OF TIMES IF
YOU LIKE – ACHILLES NEVER
CATCHES THE TORTOISE.

It's easy to see the relationship of this paradox to the number series if Achilles runs twice as fast as the tortoise (perhaps he's damaged his Achilles tendon, or the tortoise is on steroids).

In the time Achilles covers the first distance, the tortoise moves half that distance. While Achilles is catching up, the tortoise moves ¼ the original distance. In an infinite set of moves they will only get to twice the original distance (which is where, of course, the paradox falls down as Achilles powers through that mark). This paradox is the infinite series $1 + \frac{1}{2} + \frac{1}{4} \ldots = 2$.

CALL YOURSELF A HERO?

Apeiron

There's something unsettling, both about the idea of infinity itself and about the way that these infinite series can have a finite sum. The Greeks weren't sure what to do with the concept of infinity. They called it *apeiron* (roughly pronounced a-pay-a-ron). For us "infinity" is a fairly neutral word – if anything it has grand and dramatic associations. But for the Greeks, *apeiron* had the same sort of negative connotations that "chaos" does today. In a culture that placed a great deal of emphasis on precision, *apeiron* was indefinite and immeasurable.

APEIRON IS UNBOUNDED, UNCONTROLLED AND DANGEROUS.

Aristotle

The man who dealt with infinity in a way that satisfied the Greeks, and most mathematicians until the 19th century, was **Aristotle**. Born in 384 BC in Stagirus, Aristotle joined Plato's Academy. This was not just *an* academy, it was *the* academy, the original, set up in the grove of a man named Academos.

Aristotle, in the classic armchair-musing fashion of a Greek philosopher, examined infinity. He began by looking at existing views. The Pythagoreans thought infinity was "what was outside heaven", while the atomists believed it was the attribute of a substance, like its colour, rather than something that existed in its own right.

IT IS INCUMBENT ON THE PERSON WHO SPECIALIZES IN PHYSICS TO DISCUSS THE INFINITE AND TO INQUIRE WHETHER THERE IS SUCH A THING OR NOT, AND, IF THERE IS, WHAT IT IS.

Infinity had to exist, Aristotle decided, because time did not have a beginning or an end. Nor did the counting numbers stop. And the universe could be without limit. Yet infinity was no more something in its own right, as the Pythagoreans thought, than was number or magnitude. Worse still, infinity also *couldn't* exist. Aristotle's arguments are obscure, but the clearest one imagines an infinite body. It would have to be unbounded, because otherwise it would be finite. Yet a body is defined by its bounds – that is how you distinguish it from everything else. So an infinite body can't exist.

IN VIEW OF THE ABOVE CONSIDERATIONS, NEITHER ALTERNATIVE SEEMS POSSIBLE ... AND CLEARLY THERE IS A SENSE IN WHICH THE INFINITE EXISTS AND ANOTHER IN WHICH IT DOES NOT.

WHERE IS IT THEN?

Potential infinity

Aristotle decided that infinity was a **potential** state. This might be difficult to grasp, but Aristotle gave us a beautiful picture to understand it.

Think, he said, of the Olympic Games. They exist – no one can doubt that. But imagine that a little green man comes down in a flying saucer (that's my extension to the story) and says: "Show me this Olympic Games of which you speak? Where is it?". Well, it's not there. I can't show you it. The Olympic Games exist, but they aren't something we can point to (except for two weeks every four years). They are potential, just as infinity is potential.

Left brain/right brain

When we look back at the ancient Greeks it's easy to misunderstand their thinking. Where our maths is **representational**, using symbols to stand in for values, theirs was much more **visual** – hence the enthusiasm for geometry. The Greeks had a mathematics of imagery.

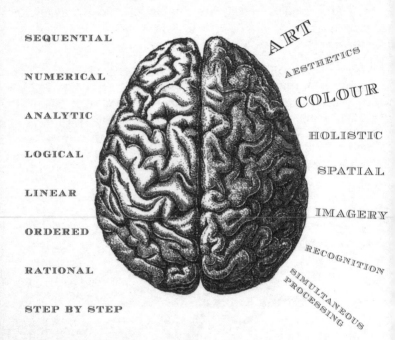

SEQUENTIAL

NUMERICAL

ANALYTIC

LOGICAL

LINEAR

ORDERED

RATIONAL

STEP BY STEP

ART

AESTHETICS

COLOUR

HOLISTIC

SPATIAL

IMAGERY

RECOGNITION

SIMULTANEOUS PROCESSING

Studies of the brain have shown that it can operate in two different modes, usually labelled left and right brain, because one side or the other dominates. When we approach mathematics we go in left brain blazing. It's all about logic and system and number and analysis. But for the Greeks the right brain was dominant. Their mathematics was highly visual.

The power of algebra

There are many mathematical problems that we would now tackle using **algebra**. Some may have found this a nightmare at school, but it's attractive to maths fans because it's basically a puzzle.

The first step is to make the problem more compact. I could say that the amount I've got in my bank after one year is *the original amount plus that same amount times the level of interest*. But it's much easier to see what's going on (once you get over any discomfort with the symbols) to say:

N = O x (1 + i)

… where N is new amount, O is old amount and i is interest rate.

The power of algebra starts coming through when you get a little more complexity, or you have a form with a particular solution. So, for example, anyone who has done basic physics will be aware that the kinetic energy of a moving object is given by:

$$KE = \tfrac{1}{2}mv^2$$

… where m is mass and v is velocity, which would be much more fiddly in words. Similarly, most of us will have suffered quadratic equations at school and may vaguely remember that the solutions to:

$$ax^2 + bx + c = 0$$

Will always be:

$$\frac{-b \pm \sqrt{(b^2 - 4ac)}}{2a}$$

… even if we aren't sure why we should care.

In a way it's a good job that the Greeks didn't go in for algebra, because their equations would have been a nightmare. Imagine the simple formula: A + B = C + D. The Greeks had none of the operator symbols that keep this concise. They would have to write out the whole thing in text. And to make matters worse, they wouldn't have bothered with spaces between words.

Their formula would have looked something like this:

THEAANDTHEBTAKENTO-
GETHERAREEQUALTO-
THECANDTHEDTAKENTO-
GETHER

Yet for them this wasn't the problem we would see it to be, because they would approach many of the mathematical challenges that we would assign to algebra from a visual viewpoint.

Visual thinking

Another complication is that the Greeks didn't use fractions. Instead of "a half" they said "the second part". This is visual thinking. Instead of considering something to be half the size of another, they would think of a shape that fits into another one twice. Parts were defined by the number symbol with a dash over it – letters were used as numbers, so gamma (γ) was 3, while "the third part" was gamma with a dash. Confusingly, beta with a dash was 2/3, with a special symbol for "the second part".

The lack of fractions made arithmetic tricky – you needed a book of tables to add parts.

This visual approach to maths favoured by the ancient Greeks could sometimes be enlightening. The discovery that the series $1 + \frac{1}{2} + \frac{1}{4} \ldots$ adds up to 2 is quite unnerving when you think of it as adding an infinite series of numbers together and coming to a finite value. It doesn't seem right that an infinite set of *anything* should make just 2. If you visualize it, though, it looks reasonable.

Starting with a unit shape, add in the second part, the fourth part, the eighth part, the sixteenth part and so on. Approached visually, it's clear that you will never fill the box without adding an infinite set of parts.

Pythagorean perfection

Even so, there were some visually derived fractions that were anything but natural to the Greeks. There's no better example than the affair of the Pythagoreans and the diagonal of a square.

Pythagoras, another early Greek mathematician born in 569 BC, had a school that didn't just study numbers but equated whole numbers with *creation*. The universe was thought to be built on whole numbers and their ratios. Each number from 1 to 10 was considered to have vital symbolism.

1 IS THE *UNIQUE MIND*. 2 REPRESENTS *OPINION*, WHICH IMPLIES CONVERSATION. 3 IS *WHOLENESS* (NEEDING A BEGINNING, MIDDLE AND END). AND SO ON.

10 was a very special number, the number of perfection. 10 is the sum of the first four numbers, but thinking visually like the Greeks, when objects are arranged in the series 1–4 above each other they form a perfect triangle, the simplest of the shapes.

GOD IS NUMBER

The Pythagoreans were so convinced that 10 was central to creation that they insisted there had to be an unknown tenth heavenly body, an anti-Earth that was always behind the Sun. Odd numbers were male, even numbers female. And being the Pythagorean crew, they had quite an interest in such matters as diagonals of a square.

Diagonals of a square

Let's keep it simple and think of a square that's one unit on each side. How long is that square's diagonal? We know from Pythagoras' theorem that all we have to do is multiply each of the two sides by itself, add them together and find the square root. Here that's 1 x 1 and 1 x 1 again, making 2. Take the square root of this – the number that is multiplied by itself to make 2 – and we have the answer.

IT'S OBVIOUSLY BIGGER THAN 1 AND LESS THAN 2. WE KNOW THAT IT HAS TO BE A RATIO OF TWO NUMBERS ... BUT WHAT RATIO?

To find the diagonal, the Pythagoreans were looking for a ratio of two numbers a and b where $a/b = \sqrt{2}$. With basic logic, depending only on a knowledge of odd and even numbers, it's possible to show that the square root of 2 is not *any* ratio of two whole numbers. The argument goes like this.

Imagine there's a ratio $a/b = \sqrt{2}$ and this is the simplest ratio, so a and b are the smallest integers they can be. That's the same as $a^2 = 2 \times b^2$.

AS 2 X ANYTHING IS EVEN, A^2 MUST BE EVEN, SO A IS EVEN (BECAUSE ODD X ODD = ODD).

Drowning by numbers

If a is even, it can be divided by 2. So a^2 can be divided by 4. But $a^2 = 2 \times b^2$, so b^2 can be divided by 2. So b^2 (and b) is even. Both a and b are even; they can both be divided by 2. This counters our starting point, setting a and b as small as possible. So you can't have a ratio $a/b = \sqrt{2}$. The square root of 2 is **irrational***: it can't be made from a **ratio** of whole numbers.

According to legend, the Pythagoreans were so horrified at this discovery that when one of their school, Hipparsus, let out the secret, he was forcibly drowned.

Historically, this whole business sounds unlikely. Remember, the Greeks didn't have the concept of fractions in the same way as we do. Their mathematical work was geometrical, dealing with diagrams, not formulae. Their view of the length of the diagonal of a unit square was that it was something between 1 and 2 that was not in the ratio of lengths of any two sides of an object.

In practice, the Pythagorean philosophy of the perfection of number was entirely separate from geometry, which was not directly related to numbers.

Squaring the circle

In fact, the diagonal of a unit square is a relatively tame irrational. We can easily write down a formula that describes it. But the Greeks were also aware of less tractable irrationals. The obvious example is the ratio of the circumference of a circle to its diameter. The challenge of working this out fascinated the Greeks, as did the associated problem of working out the size of square with the same area as a circle.

WHAT A TETRAGONIDZEIN.

Such was the fascination of "squaring the circle" that the Greeks had a word for someone who spent their time trying to do it – τετραγονιδζειν (*tetragonidzein*).

Transcendental pi

We now know what the problem with squaring the circle is. The irrational number at the heart of the circle is pi (π) – 3.14159… Unlike the square root of 2, this is a number that doesn't have a simple relationship to whole numbers. We can't write down a finite equation to calculate pi. Pi is the best known of the **transcendental*** numbers. Just as irrational numbers have nothing to with lacking rational explanation, a transcendental number isn't likely to engage in sudden bursts of meditation or yogic flying. The term merely says that the number *transcends* – is outside of – calculation using an equation with a finite set of terms.

Infinity of pi

Pi embodies a kind of infinity. You would have to write out an infinitely long decimal to capture its value exactly. It has now been calculated to many millions of places. For those who enjoy showing off their number skills there are a range of rhymes where the word lengths indicate the sequence of digits. Such was a little ditty sent by Adam C. Orr of Chicago to the *Literary Digest* in 1906:

3.1   4   1   5   9
NOW I – EVEN I – WOULD CELEBRATE

2   6   5   3   5
IN RHYMES UNAPT THE GREAT

8   9   8   9
IMMORTAL SYRACUSAN RIVALLED NEVERMORE,

3   2   3   8   4
WHO IN HIS WONDROUS LORE,

6   2   6
PASSED ON BEFORE,

4   3   3   8
LEFT MEN HIS GUIDANCE

3   2   7   9
HOW TO CIRCLES MENSURATE ...

This doesn't mean that it's impossible to calculate pi using a formula – such methods have been available since the 16th century. But unlike √2, the formula for pi (and other transcendental numbers) depends on the sum of an infinite series, rather than a finite equation that can be fully written down. With enough time and computing effort you can get as close as you like to pi, but you can't calculate it absolutely. The first simple formula, from Newton's contemporary John Wallis, was:

π/2 = 2/1 x 2/3 x 4/3 x 4/5 x 6/5 ...

It's a simple sequence, but it can't be fully written out.

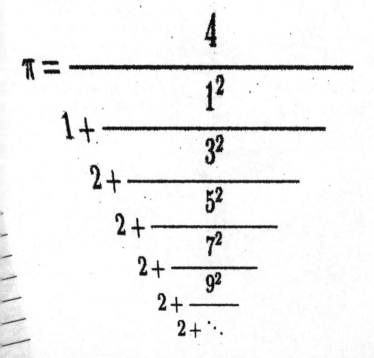

$$\pi = \cfrac{4}{1 + \cfrac{1^2}{2 + \cfrac{3^2}{2 + \cfrac{5^2}{2 + \cfrac{7^2}{2 + \cfrac{9^2}{2 + \ddots}}}}}}$$

Omega

Although transcendental numbers are often the most intractable numbers considered, some are even more incalculable. The best known is the number named "Omega" (Ω) by American mathematician Greg Chaitin. Omega is "unknowable". If you think of the other numbers in terms of computer programs to generate them, even pi can be generated by a relatively short program (it would just have to be run for an infinite time to get an exact value). But Omega cannot be produced by any program. It's a sequence of digits that has no pattern or structure. The only way of generating it is to write it out, digit by digit.

IF I COULD TELL YOU HOW TO CALCULATE IT, IT'S NOT OMEGA.

Not really numbers at all?

By the middle ages, there was a grudging acceptance that irrational numbers like the square root of 2 had to exist, but mathematicians avoided them if at all possible. The 16th-century German mathematician **Michael Stifel** (c. 1486–1567), who was one of the inventors of logarithms, and who introduced some of our best-known mathematical symbols like + and (ironically) √, acknowledged the value of irrational numbers but was at pains to say that in many senses they weren't numbers at all. These were not values that could be worked with in the normal way but rather they lay "hidden in a kind of cloud of infinity".

WE FIND THAT THEY FLEE AWAY PERPETUALLY, SO THAT NOT ONE OF THEM CAN BE APPREHENDED PRECISELY IN ITSELF.

God and the infinite

For the Christian philosophers who followed the Greeks, infinity was a topic best left to God. The bishop philosopher **Augustine of Hippo** (AD 354–430) was clear that numbers could not stop, but had to reach infinity. Some people, he suggested, think that knowledge can't encompass infinity, even if it's God's knowledge. But for Augustine, though it's impossible for us to "number the infinite", to limit God who created the concept is ridiculous. In fact, according to Augustine, the period of time before God created the universe was infinite, an eternity, compared with which the universe had been in existence for the shortest imaginable time.

> THE INFINITY OF NUMBER, THOUGH THERE BE NO NUMBERING OF INFINITE NUMBERS, IS YET NOT INCOMPREHENSIBLE BY HIM WHOSE UNDERSTANDING IS INFINITE.

Some later theologians, like the Italian monk **Thomas Aquinas** (1225–74), would not agree. Influenced by the Arab philosophers and the newly rediscovered works of Aristotle, Aquinas would argue that though God was unlimited, that did not mean that he could do the impossible. He couldn't produce a square circle or a visible object that was invisible. For Aquinas, God would have the same problems with infinity. It's not that he couldn't do anything meaningful he set his mind to, but making something infinite, or even envisaging the infinite, was not within his grasp because it had no reality as a concept.

> ALTHOUGH GOD'S POWER IS UNLIMITED, HE STILL CANNOT MAKE AN ABSOLUTELY UNLIMITED THING, NO MORE THAN HE CAN MAKE AN UNMADE THING.

The linkage of God with the infinite comes up repeatedly in world religions. In the Hindu scripture the *Bhagavad Gita*, we read: "O Lord of the universe, I see You everywhere with infinite form ..."

And in the Jewish religion that gave birth to both Christianity and Muslim traditions there are specific references to infinity in the mystical tradition of the Kabbalah. The Kabbalah is very much driven by number. At the heart of the Kabbalah are ten properties or components called the Sefirot.

THESE COMPONENTS ARE ALL CONSIDERED SUBSIDIARY TO THE GODHEAD – AND THAT IS CALLED *EIN SOF* – IN EFFECT, INFINITY.

The human perspective

For later philosophers, it was more important to consider the human take on infinity than to worry about God's abilities. Philosopher **David Hume** (1711–76) decided not only that human beings can't conceive of infinity (because our minds are finite), but also that the infinite (and particularly the infinitesimally small) could not exist. Hume demonstrated the inability to reduce things for ever by viewing a blot of ink from a distance where it was just visible.

IF YOU THEN DIVIDE THE BLOB INTO TWO, THOSE DIVIDED PARTS DISAPPEAR.

THIS SHOWS THAT THE DISTANT IMAGE HAS BECOME VISIBLY "EXTENSIONLESS".

The inkblots, he believed, reached the limit of divisibility.

Hume's argument was flawed, equating the capabilities of the senses with reality. German mathematician **David Hilbert** (1862–1943) would suggest that the thought process could not be separated from reality. This being the case, he suggested, when we think we're dealing with infinity, in fact we're just thinking of something very, very big.

WHEN WE THINK THAT WE HAVE ENCOUNTERED INFINITY IN SOME REAL SENSE WE HAVE MERELY BEEN SEDUCED INTO THINKING SO BY THE FACT THAT WE OFTEN ENCOUNTER EXTREMELY LARGE AND EXTREMELY SMALL DIMENSIONS.

Not everyone agrees. Shaughan Lavine, Associate Professor of Philosophy at the University of Arizona, points out a very simple way that anyone can envisage infinity. As long as you can grasp the meaning of "finite" and the meaning of "not", he says, you should have a simplistic picture of the infinite.

"Only a manner of speaking"

Even so, some serious mathematicians never accepted the reality of infinity, even as a mathematical concept. The great German mathematician **Johann Carl Friedrich Gauss** (1777–1855) was convinced that infinity was an illusion, like the end of the rainbow, that could never be reached, even though we can aim for it. "The infinite", said Gauss, "is only a manner of speaking, in which one properly speaks of limits to which certain ratios can come as near as desired, while others are permitted to increase without bound."

> I PROTEST AGAINST THE USE OF AN INFINITE QUANTITY AS AN ACTUAL ENTITY; THIS IS NEVER ALLOWED IN MATHEMATICS.

Yet well before Gauss's time, someone did dare to take on the possibility of a real infinity, face to face.

Galileo

When it came to dealing mathematically with infinity, the Greek view passed relatively unchanged through to the Renaissance. The first new thinking came from that remarkable challenger of the status quo, **Galileo Galilei** (1564–1642).

Galileo tends to be remembered for dropping balls off the Tower of Pisa, something he probably never did (he was a great self-publicist, but he never mentioned it – it was only recorded many years later by an assistant) and for being locked up for daring to suggest that the Earth rotates around the Sun. But he also undertook some remarkable thinking on the subject of infinity.

GALILEO GALILEI

Galileo's infinite pondering took place after his trial. Until then, his career had been going brilliantly. Although he didn't, as is often suggested, invent the telescope, he heard about its development in Holland.

WHEN A DUTCH SPECTACLE-MAKER HEADED FOR VENICE TO SHOW OFF THIS NEW DEVICE, I HAD A FRIEND DELAY MY RIVAL SO I COULD QUICKLY CONSTRUCT A TELESCOPE.

He was rewarded with a well-paid job and would have been a great success if he hadn't written his book supporting Copernicus' Sun-centred model of the universe. The way he wrote this offended the religious hierarchy; he was put on trial and sentenced to life imprisonment.

While under house arrest after his conviction, Galileo put together his masterpiece, *Discourses and Mathematical Demonstrations Concerning Two New Sciences.*

MATHEMATICAL
DISCOURSES
CONCERNING
Two New Sciences
RELATING TO
Mechanicks and Local Motion,
IN
FOUR DIALOGUES.

I. Of the Resistance of Solids against Fraction. | III. Of Local Motion, viz. Equable, and naturally Accelerate.
II. Of the Cause of their Coherence. | IV. Of Violent Motion, or of Projects.

By GALILEO GALILEI,
Chief Philosopher and Mathematician to the Grand Duke of TUSCANY.

With an APPENDIX concerning the Center of Gravity of SOLID BODIES.

Done into English from the Italian,
By THO. WESTON, late Master, and now publish'd by JOHN WESTON, present Master, of the Academy at Greenwich.

LONDON:
Printed for J. HOOKE, at the Flower-de-Luce, over-against St. Dunstan's Church in Fleet-street. M. DCC. XXX.

I HAD REAL TROUBLE GETTING THIS PUBLISHED – THE INQUISITION WERE NOT TOO KEEN AFTER MY PREVIOUS BOOK.

BUT TO MY GREAT SUR- PRISE, IT WAS EVENTUALLY TAKEN UP BY THE DUTCH PUBLISHER ELSEVIER.

The book (like his near-fatal work on the motion of the Earth) took the form of a conversation between characters. After wondering about what holds matter together, they have a diversion, just for the fun of it, into the nature of infinity.

Infinity on wheels

The most dramatic demonstration of Galileo's ideas on infinity involved wheels. He imagined a pair of multi-sided wheels, one stuck to the face of the other, running on rails. Say they're hexagons. We give them a turn until they move from one face on the rails to the next face. The bigger wheel will have moved forward by the length of one of its sides.

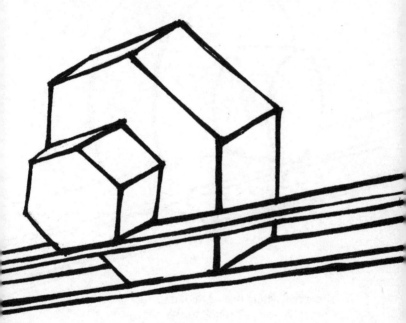

The smaller wheel has to move this distance too, even though its sides are shorter, because the wheels are fixed together. It manages to do this by lifting off its rail for just far enough to keep up with the larger wheel.

Now here's the clever bit. Galileo imagines similar wheels but with more and more sides. The more sides, the smaller the jump the small wheel has to make to catch up. At the extreme we get to two circular wheels. Say we give them a quarter-turn. They both move forward a quarter of the circumference of the big wheel.

The small wheel has never left its rail – yet it has moved much further than a quarter of its circumference. Galileo argued that it manages this by making an infinite set of infinitely small jumps to bridge the gap.

One of the characters in the book, Simplicio, is a little slow and is there to say "Duh, I don't understand", so the others can explain. After letting the circular wheels percolate through his brain, Simplicio has a complaint. Galileo seems to be saying that there are an infinite number of points on each of the wheels – but somehow, one infinity is bigger than the other. The response is rueful. That's just the way it is with infinity – a problem, Galileo reckons, of dealing with infinite quantities using finite minds. And he goes on to show how this is perfectly normal behaviour for the infinite.

REMEMBER THAT WE ARE DEALING WITH INFINITIES AND INDIVISIBLES, BOTH OF WHICH TRANSCEND OUR FINITE UNDER-STANDING ... IN SPITE OF THIS, MEN CANNOT REFRAIN FROM DISCUSSING THEM.

Back to geometry

One way Galileo demonstrates the odd mathematics of infinity is to use geometry, the favourite tool of the ancient Greeks. He gives a geometrical proof that you can design a cone and a bowl (the latter carved from a solid slice of a cylinder) in such a way that when you make a straight line cut through them, each has the same area and volume at the point of the cut, at whatever level you make that cut. Yet if we move the cut to the top, we appear to have a point and a circle, both proved to be the same "size".

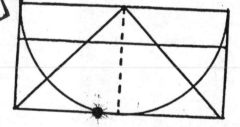

THIS PRESENTATION STRIKES ME AS SO CLEVER AND NOVEL THAT, EVEN IF I WERE ABLE, I WOULD NOT BE WILLING TO OPPOSE IT.

The normal rules don't apply

Simplicio doesn't feel this helps, so Galileo tries again. He makes sure Simplicio knows what a square is – any number multiplied by itself. Then he imagines going through the real numbers, multiplying each by itself. For every positive integer there is a square. We've an infinite set of integers, and there's a square for each integer. But here's the rub. There are far more integers than there are squares.

$$1, 2, 3, 4, 5, 6,$$
$$7, 8, 9, 10, 11, 12, 13, 14, 15, 16, 17,$$
$$18, 19, 20, 21, 22, 23, 24, 25, 26,$$
$$27, 28, 29, 30, 31, 32, 33, 34, 35, 36$$

Simplicio's brain hurts, and it doesn't surprise us. Galileo has spotted something special about infinity. The normal rules of arithmetic don't apply. You can have "smaller" and "bigger" infinities that are the same size.

> WE ARE LED TO CONCLUDE THAT THE ATTRIBUTES "LARGER", "SMALLER", AND "EQUAL" HAVE NO PLACE IN COMPARING INFINITE QUANTITIES ...

The infinity of 1?

This odd arithmetic of infinity led Galileo to a strange (and erroneous) conclusion. His argument went something like this. There have to be as many squares as there are natural numbers*. But the bigger the number, the more sparsely distributed are the squares (there are many more numbers that aren't squares). So the bigger the number gets, the further you get from infinity. Since the further you get down the number line* the further you are from infinity, it follows that by turning back we discover that if any number is truly infinite it is 1. And just like infinity, 1 x 1 = 1.

> I MEAN THAT UNITY CONTAINS IN ITSELF AS MANY SQUARES AS THERE ARE CUBES AND NATURAL NUMBERS.

A common error

The trap Galileo seems to have fallen into is a common one – the assumption that when two things have similar properties they can be equated. This is part of the basis of homeopathy, where the idea of the "law of similars" is that a poison that causes similar symptoms to a disease will cure that disease. Galileo equated infinity and unity because they had similar properties.

UNFORTUNATELY, IF YOU TOOK GALILEO'S ARGUMENT LITERALLY YOU WOULD THINK THAT EVERYTHING WITH WHITE CURLY FUR AND FOUR LEGS WAS A SHEEP – EVEN IF WHAT YOU WERE LOOKING AT WAS A WHITE SOFA WITH A FURRY COVER.

The indivisibles

Around this time, the idea of **indivisibles** became popular. This was a similar approach to the ancient Greek idea of atoms. Ancient Greek atoms were the result of cutting something up so small that it was no longer possible to cut any further (*a-tomos* means "uncuttable"). The use of indivisibles involved dividing an object into smaller and smaller pieces, but not necessarily in all three dimensions. Take the example of the area of a circle.

IMAGINE DIVIDING THE CIRCLE INTO LOTS OF THIN SEGMENTS, LIKE PIECES OF AN ORANGE.

This reflected an idea that went back as far as the ancient Greek philosopher Antiphon.

Antiphon was a contemporary of Socrates, born in the 5th century BC. He suggested that you could work out the area of a circle by drawing a regular polygon inside it and gradually increasing the number of sides until it was closer and closer to the circle itself. But 15th-century philosopher Nicholas of Cusa went further. He imagined stacking segments of a circle on top of each other, alternating direction. This made something very close to a rectangle that would be πr in height and r in width, making its area πr^2. Of course the edges of those segments would never be quite straight unless the indivisibles were *infinitely narrow*.

VOILA!

61

Newton and potential infinity

Galileo took a peek under the carpet at the "real" infinity and found it entertainingly baffling. But before this true infinity was taken further, a war broke out over its virtual counterpart, Aristotle's "potential infinity". This wasn't a war between nations but in the mathematical world. The first contender in the fight was **Isaac Newton** (1642–1727). It's often said that Newton was born the same year that Galileo died, as if accepting the baton of greatness. Paradoxically, this is both true and false. In the old dating system, Newton was born on Christmas Day 1642 – but in the modern calendar it was January 1643.

Newton was a remarkable man. He had many achievements, from his theories of light and colour and his exploration of the concept of gravity, to predicting the celestial mechanics of planetary motion. And he also developed the mathematics needed to deal with such complex motions – a trick that depended on infinity, which he seems to have invented early in his career but did not communicate until much later.

> THE TROUBLE WITH MOTION DUE TO GRAVITY AND SIMILAR PHYSICAL PROBLEMS IS THAT THEY INVOLVE **ACCELERATION**.

> BODIES AREN'T MOVING ALONG SMOOTHLY, BUT CHANGING IN SPEED ALL THE TIME. THERE HAS TO BE SOME WAY TO DEAL WITH THESE CHANGES.

Fluxions

The mathematical trick, which Newton called the method of **fluxions**, was designed to help with the calculation of values useful in dealing with acceleration – the rate at which something is changing. This is easy to do when the acceleration is *linear*. Imagine a car accelerating from a standstill, and, extremely conveniently, the speed increases with time as a nice straight line.

THEN THE ACCELERATION – THE RATE AT WHICH THE SPEED CHANGES – IS JUST THE SLOPE OF THAT LINE: THE CHANGE IN SPEED DIVIDED BY THE CHANGE IN TIME. IT'S LIKE WORKING OUT THE GRADIENT OF A HILL.

What Newton realized was that, when dealing with acceleration that follows a *curve*, if you make the change small enough, zooming in to the detail of the curve until you're practically dealing with a point, then to all intents and purposes here again is a straight line. And so for that tiny little section of the curve it's almost exactly true that the acceleration is the change in distance divided by the change in time.

> BY CONCENTRATING ON JUST A MINUSCULE SEGMENT OF THE CURVE, WE CAN DEAL WITH IT AS IF IT'S A STRAIGHT LINE, MAKING IT ACCESSIBLE TO STRAIGHTFORWARD MATHS.

SPEED.

0

TIME.

From *o* to 0

Newton called the rate at which a quantity changed a **fluxion** and the value that was changing a **fluent**. He represented the infinitesimal change by a little italic *o*. Now the clever thing was, he imagined this *o* getting smaller and smaller until it was zero. As *o* got smaller and smaller the result became closer and closer to correct, until *o* vanished away and out came exactly the right result. Every time.

Newton was an odd man. He sat on this idea for many years, though he did describe it to friends.

HE ALSO MENTIONED HIS RESULTS WITHOUT ANY REASONING IN A LETTER TO THE GREAT GERMAN MATHEMATICIAN, *GOTTFRIED WILHELM LEIBNIZ* (1646–1716).

Dear Gottfried, How you doing? Got something you might fin

The closest Newton came to an explanation in his letter was an obscure coded remark. It represents: "Given an equation that consists of any number of flowing quantities, to find the fluxions: and vice versa"; or to be more precise, the Latin version of it, in the form of a count of each letter present. (There are a lot of v's, as u's and v's weren't distinguished in Latin: *Data æqvatione qvotcvnqve flventes qvantitates involvente, flvxiones invenire: et vice versa*.)

I cannot proceed with the explanation of the fluxions now, I have preferred to conceal it thus 6accdæ13eff7i3l-9n4o4qrr4s8t12vx ...

Beyond such cryptic utterances, Newton wouldn't publish his ideas for more than 30 years. But this doesn't reduce the importance of fluxions. They transformed the mathematics of movement, providing the ultimate new technology of the time.

Leibniz's calculus

Leibniz did much more than read Newton's letters. He too was working on the problem – as far as we can tell, quite independently. He came up with his method after Newton, but published first. And he called it **calculus***, using the notation still in use today.

> THOUGH BASED ON THE SAME MATHEMATICS, MY NOTATION AND TERMINOLOGY WAS EASIER TO MANAGE.

His calculus soon became popular, leaving Newton incandescent with rage. He had no tolerance for competition. Newton accused Leibniz of plagiarism, and would continue to consider that Leibniz had robbed him of glory for the rest of his life – and that was another 50 years.

> GGGGGGGGG GGGGGGGGG GGGGGGGGG GGGGRRRRR RRRRRRRRR RRRRRRRR.

Newton vs. Leibniz

Leibniz felt slighted. Apart from anything else, he had clearly published first. His paper on calculus went to print in 1684. Though Newton seems to have devised fluxions as early as 1671, they didn't appear in print until 1687. Meanwhile, accusations, voiced in the icy politeness of the day, flew backwards and forwards across the Channel. Leibniz felt forced to act when one of Newton's friends, John Keill, published a Royal Society paper making explicit accusations of plagiarism.

F⊘⒢'@*?
$#☠!

TRANSACTIONS OF THE
ROYAL SOCIETY, 1708

Given the way
Newton generously
shared his ideas
with Leibniz, there
is every possibility
that Leibniz simply
applied his own
terminology
to Newton's
invention.

Leibniz was wounded. He was a Fellow of the Royal Society and didn't expect such treatment. Keill was asked to apologize, but only acknowledged that Leibniz's notation was original – hardly an apology. Leibniz complained again. This pushed the Royal Society into action. A committee of eleven men was set up to establish the truth. The final report was written by no less a figure than the Society's president. Surely this would appease Leibniz?

SADLY NOT. THE PRESIDENT OF THE ROYAL SOCIETY, THE AUTHOR OF THIS "IMPARTIAL" REPORT, WAS SIR ISAAC NEWTON HIMSELF!

The rift between British and continental mathematicians lasted a century.

Notation

One reason why Leibniz's calculus has proved more popular than Newton's fluxions is that the notation Leibniz invented was so much more practical than Newton's. Newton's squashed o caused confusion with zero, and didn't give any information about what was changing.

> BY CONTRAST, I BUILT ON EXISTING NOTATION TO PRODUCE A SYSTEM THAT WAS EASIER TO FOLLOW AND MUCH EASIER TO MANIPULATE.

There had been a convention for some time to use the capital Greek letter delta (Δ) to indicate change. A small delta (δ) meant a small change, so for instance, δx meant a small change in the quantity x.

Leibniz went one step further from the use of δx to mean a small change in x, instead writing dx to mean an infinitesimally small change in x, the change needed to perform the trick of regarding a curve as a straight line. Where the change in x was dx, the rate of change of x was dx/dt, where t is time.

I HAVE AN EQUIVALENT SYMBOL FOR THE RATE OF CHANGE OF X IN MY NOTATION: AN X WITH A DOT OVER IT, REFERRED TO AS "PRICKED NOTATION".

THE DOTS ARE EASY TO MISS AND LESS INFORMATIVE THAN MY ALTERNATIVE.

Differential and integral calculus

As well as dealing with rates of change and similar problems – so-called **differential calculus** – both men's technique handled the kind of thing Nicholas of Cusa had done when he piled up the segments of a circle to calculate an area. This **integral calculus** is used to find the area under curves, the volumes of three-dimensional objects and so on. Newton simply regarded integration* as the reverse of differentiation* (technically true) and had no special symbols. Leibniz regarded it as a summing process and used a stretched version of the letter S (for "summa") to produce the integral sign ∫ that we use today.

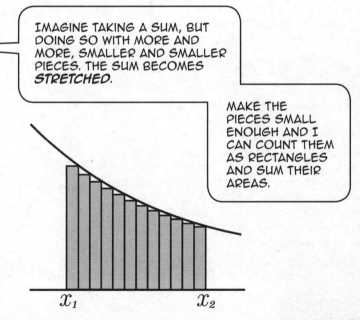

IMAGINE TAKING A SUM, BUT DOING SO WITH MORE AND MORE, SMALLER AND SMALLER PIECES. THE SUM BECOMES *STRETCHED*.

MAKE THE PIECES SMALL ENOUGH AND I CAN COUNT THEM AS RECTANGLES AND SUM THEIR AREAS.

x_1 x_2

Battling Bishop Berkeley

Yet impressive though the sparks flying between the two were, Newton and Leibniz were not the sole contributors to the fluxion wars. For completeness, we need to introduce **Bishop George Berkeley** (1685–1753) who disagreed with both. Berkeley is best remembered as the philosopher who asked whether or not an event occurs without an observer. For example, if a tree falls in the middle of a forest with no one aware of it or there to observe it, can it be said to happen? Berkeley answered this with something of a fudge.

GOD IS ALWAYS PRESENT EVERYWHERE SO THE TREE WILL BE OBSERVED AND DOES FALL.

Berkeley certainly wasn't an old fogey naturally opposed to the scientific developments of the day as, say, the bishops opposed to evolution were later portrayed. Significantly younger than Newton, he was quite a character. Before settling down as Bishop of Cloyne in Ireland he had spent some time in the Americas.

> I BOUGHT A HOUSE IN NEWPORT, RHODE ISLAND AS A BASE IN AN ATTEMPT TO SET UP A COLLEGE IN THE WEST INDIES. BUT FUNDS RAN OUT AND I HAD TO RETURN HOME.

Berkeley published a stinging attack on fluxions (and calculus for that matter) under the magnificent title, *The Analyst: A discourse addressed to an infidel mathematician.*

THE

ANALYST;

OR, A

DISCOURSE

Addressed to an

Infidel MATHEMATICIAN.

WHEREIN

It is examined whether the Object, Principles, and Inferences of the modern Analysis are more distinctly conceived, or more evidently deduced, than Religious Mysteries and Points of Faith.

By the AUTHOR of *The Minute Philosopher.*

The SECOND EDITION,

First cast out the beam out of thine own Eye; and then shalt thou see clearly to cast out the mote out of thy brother's eye. S. Matt. c. vii. v. 5.

LONDON:

Printed for J. and R. TONSON and S. DRAPER

The infidel mathematician

Oddly, Bishop Berkeley produced his argument in response not to Newton or Leibniz, but **Edmond Halley** (1656–1742), the Astronomer Royal, the "infidel mathematician" of the title. Halley was a vocal atheist and had persuaded a friend of Berkeley's to denounce Christianity on his deathbed. Berkeley found this deeply offensive, and this led to his outburst on fluxions. Halley, both Astronomer Royal and Savillian professor of geometry at Oxford, was a great supporter of Newton. He had personally published Newton's great work *Principia* to make sure it reached the public. And inevitably he was all for fluxions. But Berkeley found something inconsistent, almost hypocritical about this.

THOUGH I AM A STRANGER TO YOUR PERSON, YET I AM NOT, SIR, A STRANGER ... TO THE AUTHORITY THAT YOU THEREFORE ASSUME IN THINGS FOREIGN TO YOUR PROFESSION; NOR TO THE ABUSE THAT YOU, AND TOO MANY MORE OF THE LIKE CHARACTER, ARE KNOWN TO MAKE ...

Berkeley pointed out that the method of fluxions involved an inconceivably small quantity – one so small that it was, in effect, zero, but that still had a value. Berkeley referred to Newton's disappearing o's as *the ghosts of departed quantities*. This could only be taken on trust. Fair enough. But Halley regularly attacked Christianity specifically because it required **faith**: it could not be submitted to absolute proof.

HYPOCRISY!

This wasn't just nit-picking. There *is* something worrying at the heart of fluxions (and just in case anyone tried to wriggle out this way, Berkeley also points out that Leibniz's calculus has the same problem).

Dividing zero by zero

Let's imagine we're dealing with a simply accelerating space-craft. Its speed at any time is the square of the length of time it has been flying. The acceleration is the slope of the curve. The change in time is o and the change in speed (remember speed is time squared) is $(time + o)^2 - time^2$.

Expanding that out, the change in speed is $time^2 + 2 \times time \times o + o^2 - time^2$, which is $2 \times time \times o + o^2$. We then divide it by the change in time to get the slope, which gives us:

$(2 \times time \times o + o^2)/o$

Newton cancelled out the squashed o to get to:

$2 \times time + o$

Finally he let the o ebb away to nothing. And the result was 2 x time – which is correct. But notice what he did. The fact that o became zero has a big impact on the previous step, where the o on top and bottom were cancelled out. Newton divided zero by zero.

Once you divide zero by zero, all bets are off.
Anything with zero on top should be zero.
Anything divided by zero should be infinite.
To see the confusion caused, you only have
to look at the attempts of two early Indian
mathematicians to explain this ratio.

Brahmagupta (AD 598–668)

Bhāskara (1114–1185)

ZERO DIVIDED BY
ZERO IS ZERO!

NO! ANYTHING
DIVIDED BY ZERO,
INCLUDING 0/0,
IS INFINITE.

In practice, 0 over 0 is
indeterminate – it doesn't
have a result. It's the
equivalent of asking:
"What happens when
an irresistible force is
applied to an immovable
object?" It's meaningless.

Flow and change

Newton himself didn't have a problem with this, as he saw fluxions very differently to the way we normally think of calculus. The normal process, based on Leibniz's approach, is to imagine a quantity that you shrink smaller and smaller until it approaches zero. But Newton wasn't dealing with quantities. His imagery was all about **flow**. His squashed *o* was being diluted – it was in the process of flowing away to nothing, like a sink of water emptying down the plughole.

> I AM NOT CONSIDERING WHAT *IS*, BUT HOW THINGS *MOVE* AND *CHANGE*.

For Newton, the method of fluxions was about movement, not absolute values.

Tending towards zero

At the time, despite Bishop Berkeley's efforts, the problem with calculus was mostly swept under the carpet. After all, it *worked*. Newton saw his *o* as flowing towards 0 but never reaching it – later mathematicians would say it "tended towards 0". Eventually calculus was tidied up by using a value that's just as small as you need it to be. Instead of dealing with the *infinitesimally* small, you are now dealing with something that is *inexhaustibly* small but never reaches zero. Infinity is the limit of the process, but it's a limit you never need reach. Technically it's still a fudge, but it works. Calculus does the job.

IT'S OKAY, IT'S STILL THERE.

Finding a symbol

Up to now, infinity was useless. But with calculus, infinity came into its own. Whether you were doing differential calculus and considering infinitesimally small increments, or integral calculus and adding together an infinite set of infinitely narrow segments, infinity was a working tool of the mathematician. And that meant it needed a symbol. As it happens, one had just been produced. The lemniscate, ∞, that drunken figure of eight now used to represent infinity, was introduced in a work on conic sections by **John Wallis** (1616–1703), the man behind the formula for π, who would be significantly more famous today if he hadn't had so many glorious colleagues.

LET ∞ DENOTE INFINITY.

Wallis was originally a theologian, but he turned out to be a superb code-cracker for the Parliamentarians during the English Civil War. It was this that earned him the Savillian chair of geometry as Halley's predecessor – and he was talented enough to keep the position after the restoration of the monarchy.

Wallis didn't explain why, but he casually remarked: "let the symbol ∞ denote Infinity".

PERHAPS IT CAME FROM THE OLD ROMAN SIGN FOR 1,000 (LATER REPLACED BY M) ...

... OR IT MAY BE A VARIANT ON THE SMALL GREEK LETTER OMEGA – OR JUST AN INFINITE LOOP.

But now, at least, infinity had a symbol.

The Möbius strip and Klein bottle

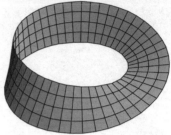

Even a circle is, in effect, an infinite loop as it has no end, but a **Möbius strip** is a much more appealing representation of infinity. Made by giving a strip of paper a single twist, then joining the two ends, the Möbius strip is a two-dimensional object that has only one side. It's easy enough to demonstrate this – put a pen on the strip at any point and draw along the paper. Eventually the pen will return to the original point, running along the whole surface. More dramatic still is the **Klein bottle** – a three-dimensional object with one surface.

A TRUE KLEIN BOTTLE WOULD HAVE TO BE TWISTED THROUGH A FOURTH DIMENSION, BUT IT'S POSSIBLE TO MAKE A 3D MODEL.

Bolzano and real infinity

Apart from Galileo, pretty well everyone since Aristotle had been dealing with potential infinity – and that's what the curve of the lemniscate represents. But in the early 19th century, Italian mathematician **Bernard Bolzano** (1781–1848) would try to get his mind around the real thing. In his retirement, and published after his death, he wrote a book called *Paradoxes of the Infinite*. He argued against philosophers like Hegel who had suggested that infinity is not a true value but rather a direction. Bolzano said that a truly infinite quantity – such as the length of a straight line unbounded in either direction – can be established.

MOST OF THE PARA-
DOXICAL STATEMENTS
ENCOUNTERED IN THE
MATHEMATICAL DOMAIN
… ARE PROPOSITIONS
WHICH EITHER IMME-
DIATELY CONTAIN THE
IDEA OF THE INFINITE,
OR AT LEAST IN SOME
WAY OR OTHER DEPEND
UPON THAT IDEA FOR
THEIR ATTEMPTED
PROOF.

Bolzano also showed something that would be very significant when another mathematician, Georg Cantor, got to work. Galileo had used the fact that every positive integer has a corresponding square to demonstrate some of the strange properties of infinity.

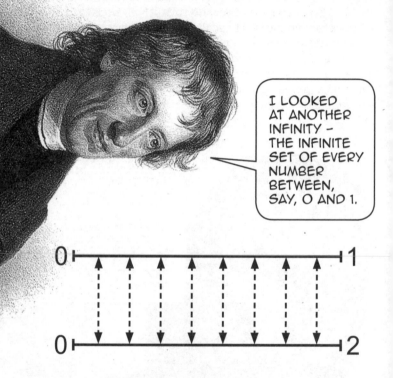

I LOOKED AT ANOTHER INFINITY – THE INFINITE SET OF EVERY NUMBER BETWEEN, SAY, 0 AND 1.

Without being able to say anything directly about the nature of this continuum of numbers, he was able to prove that it was possible to match off every number between 0 and 1 with every number between 0 and 2, just as the integers could be matched off with their squares.

Like Galileo, Bolzano had political problems. His career started with stellar promotion. In 1805, aged 24, he was awarded the chair of philosophy of religion in Prague. Although his significant work was mathematical, he was a priest and philosopher. It has been suggested that his retirement in 1820 was forced on him by the Church, but more likely it was the Viennese government. Universities still mostly insisted on teaching standard texts, but Bolzano defied the authorities by devising his own course. Worse, he preached against waging war.

MUCH OF HIS BEST WORK WAS UNDERTAKEN IN "RETIREMENT", FINANCIALLY SECURE THANKS TO A MYSTERIOUS BENEFACTOR, FRAU HOFFMAN.

Cantor: mind-bending infinity

Towards the end of the 19th century, one man used Bolzano's work as a stepping stone to think about the true nature of infinity. His name was **Georg Cantor** (1845–1918), and he would go mad as a result.

Cantor spent all his working life at the university in Halle. This is a German town famous for music, but not for maths. Cantor thought he would soon move on – and he probably would have done, had he not come up with some conclusions that were so mind-bending that at least one mathematician would set out systematically to ruin Cantor's career.

Welcome To

Halle

Birthplace of
Georg Friedrich
Händel

The joy of sets

Cantor's first great contribution was to formalize the mathematics of **sets***. Sets had been around really as long as people conceptualized – but Cantor embedded them firmly into mathematics. A set is just a group of things. They could have something in common – like the set of things that look like an orange, or the set of people with the name Brian – or they could be as disparate as the set of things you thought about today.

Cantor built on existing work to pull together a picture of how sets operate that lets us do everything from define the numbers to establish the basic mathematical operations.

> BY A SET WE ARE TO UNDERSTAND ANY COLLECTION INTO A WHOLE M OF DEFINITE AND SEPARATE OBJECTS m OF OUR INTUITION OR OUR THOUGHT.

Sets come more naturally in some languages than in others. Where sets are an afterthought in English, Chinese, for example, has sets at its heart. Descriptions in Chinese begin with the largest set, then work down through the subsets*. So, for example, an address will begin with the country, then the province, the city, the district, the street and the building. This logic extends to names too, hence the practice of putting the surname (the family clan set) before the personal name. Rather than say "Chinese people" or "British people", the Chinese set approach uses "Britain country people" (*ying guo ren*) or "China country people" (*zhong guo ren*).

> SOME OF THE OLDER CHINESE SETS ARE QUITE POETIC – FOR EXAMPLE, THERE WAS A SET OF "THINGS THAT LOOK LIKE A FLY WHEN SEEN FROM A DISTANCE".

Venn diagrams

The interaction of sets is often demonstrated using Venn diagrams. Although apparently simplistic, these images devised by Cambridge mathematician **John Venn** (1834–1923) – whose other great claim to fame was building an automated cricket bowling machine that bowled out the top Australian batsmen in 1909 – can cram in a surprising amount of detail. With just a couple of circles and a rectangle we can identify the relationships between the sets of all vehicles, cars, all red vehicles, red cars, cars that aren't red, red vehicles that aren't cars, vehicles that are red, cars or both, and vehicles that are neither cars nor red.

Venn diagrams caused bitterness in the mathematical community. The great mathematician **Leonhard Euler** (1707–83) had devised something similar, though lacking the crucial element of overlapping shapes, before Venn. But despite this heritage, Venn was attacked. In his book *The Mathematical Universe*, William Dunham writes: "No one, not even John Venn's best friend, would argue that his underlying idea is very deep … the Venn diagram is neither profound nor original. It is merely famous. Somehow within the realm of mathematics, John Venn's has become a household name. No one in the long history of mathematics ever became better known for less. There is really nothing more to be said."

Boolean algebra

A less visual way to manipulate the contents of sets is Boolean algebra, named after English mathematician **George Boole** (1815–64), whose natural talent for mathematics earned him a university chair without any formal training. Boolean algebra became hugely useful when computers came along – it's the approach taken by online search engines. Boolean algebra uses simple terms to operate on sets. For example, "AND" is the equivalent of two overlapping regions in a Venn diagram.

SO A "RED CARS" REGION IS THE COMBINATION OF "RED VEHICLES" *AND* "CARS".

IN SET THEORY THIS IS CALLED THE *INTERSECTION* OF THE SETS, JUST AS THE SHAPES INTERSECT ON THE DIAGRAM.

Another powerful Boolean term is "OR". This refers to items that might be in either of a pair of sets, effectively combining sets. On a Venn diagram, OR is the equivalent of merging two shapes. So by using "red vehicles" OR "cars", we're dealing with a set that includes all cars and all red vehicles. Because of the way it acts, this operation is called the **union** of the sets. A final important Boolean term is "NOT", which enables us to take a chunk out of a set.

SO, FOR EXAMPLE, "RED VEHICLES" *NOT* "CARS" WOULD BE THE RED VEHICLE SHAPE WITH THE CARS SHAPE CHOPPED OUT OF IT.

Making sets of the world

It's easy to think of set theory as an abstract mathematical concept, but it's central to the way we deal with the world. Technically a person is a collection of atoms, or at a less detailed level a collection of cells. But we deal with a person as a whole – the set that is "a person". Similarly, we could never interact with the world if we didn't apply sets all the time. We would have to give each animal we ever met a separate name – but instead we devise the set of dogs (say) and so can identify a particular animal as a dog.

IF WE DIDN'T TAKE THIS APPROACH WE WOULD HAVE TO LEARN HOW TO OPERATE EVERY SINGLE LIGHT SWITCH WE EVER CAME ACROSS SEPARATELY.

For our purposes we need to pick out one aspect of set theory, called **cardinality***. Let's think of two very simple sets. The first is the set of legs on my dog. The second is the four horsemen of the apocalypse. These two sets have the same cardinality if I can pair off the members of the sets on a one-to-one basis. So, for example, the front right leg could pair up with Death, front left with Famine, and so on. I've exhausted both legs and horsemen, so they have the same cardinality. But – and here's the clever thing – I needn't know how many legs or horse-men there are.

I DO KNOW IT'S FOUR, BUT THE IMPORTANT THING IS I DIDN'T *NEED* TO KNOW.

Peano and the cardinals

Before Cantor got involved, Italian mathematician **Giuseppe Peano** (1858–1932) had already used the cardinality of a set to define the cardinal numbers* – the counting numbers. Peano had some quaint notions. Medieval scholars wrote their papers in Latin for wider understanding – and so did Peano, even though Latin wasn't exactly popular by the 1890s. Later, in 1903, he devised an artificial language, *Latino sine flexione*. It was supposed to be a universal academic language to return to the easy communication that medieval universities enjoyed. His invention was simplified Latin with modern words from Italian, English, German and French added.

I PUBLISHED MY MASTERPIECE *FORMULARIO MATHEMATICO* IN *LATINE SINE FLEXIONE*, BUT THE LANGUAGE NEVER TOOK OFF.

We use the counting numbers without thinking, but Peano gave them a formal basis. Numbers have no physical reality – I can't paint a picture of five, only the symbol, or five objects. But Peano defined the cardinals using sets. He started with the empty set*, representing nothingness, the absence of anything, and defined zero as this empty set. He then defined the cardinal number of a set as the number of sets it contains, building them up like Russian *matryoshka* dolls.

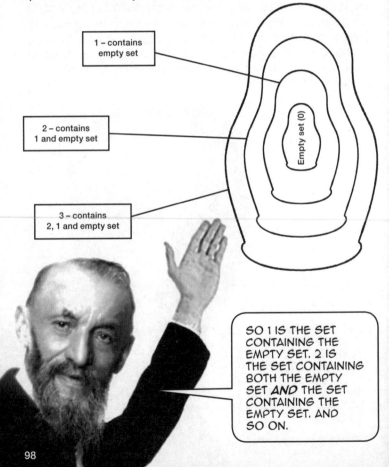

1 – contains empty set

2 – contains 1 and empty set

3 – contains 2, 1 and empty set

Empty set (0)

SO 1 IS THE SET CONTAINING THE EMPTY SET. 2 IS THE SET CONTAINING BOTH THE EMPTY SET *AND* THE SET CONTAINING THE EMPTY SET. AND SO ON.

Russell's paradox

Although set theory is the basis of much of maths, not all mathematicians are comfortable with it, because it tends to generate paradoxes. The central paradox of set theory was identified by British philosopher **Bertrand Russell** (1872–1970), who spent many years working on the philosophy and logic of mathematics. Russell's paradox depends on the idea of sets that are members of themselves. So, for example, "Everything that is not a dog" is a set that includes itself (because the set is not a dog). But the set "All pieces of music" does not include itself, because the set is not a piece of music.

SET THEORY IS AN ILLNESS THAT IS AFFLICTING MATHEMATICS, OF WHICH IT WILL EVENTUALLY BE CURED.

Henri Poincaré (1854–1912)

Russell then looked at the set "Sets that aren't members of themselves". This set would include, for instance, the set "All pieces of music".

> IS THE SET "SETS THAT AREN'T MEMBERS OF THEMSELVES" A MEMBER OF ITSELF? (YOU MAY NEED TO READ THAT A COUPLE OF TIMES.)

If it **is** a member, then it isn't a member. If it **isn't** a member, then it's not a set that isn't a member of itself – so it should be a member. It's a bit like trying to work out if the statement "This is a lie" is true. Russell showed that this paradox was fundamental to set theory.

> I WOULDN'T BE A MEMBER OF A SET THAT WOULD HAVE ME AS A MEMBER!

Cantor and subsets

With set theory in place, Cantor was ready to build on Galileo's observations on the integers and the squares (see pages 57–8). The infinite set of counting numbers has the same cardinality as the set of squares, because we can pair them off like dog legs and horsemen. And the squares are a subset of these integers. "Subset" is a spot of set theory that has become common usage. Here, it means that all the squares are members of the set "positive integers", but they aren't the full set. Cantor used this behaviour to define an infinite set.

> AN INFINITE SET HAS A ONE-TO-ONE CORRESPONDENCE WITH A SUBSET. IT HAS THE SAME CARDINALITY AS ITS SUBSET.

This underlines a crucial aspect of cardinality. However much cardinality is described as the ability to put a set into one-to-one correspondence with another set, because cardinality is a measure of size we tend to think of cardinality as the *number* of items in a set. When I say the legs on my dog has the same cardinality as the horsemen of the apocalypse, we assume this is because there are four of each – but it's not. There are infinitely fewer squares than there are positive integers. Yet the squares and the integers have the *same* cardinality because we can pair them off, one-by-one.

FOR CARDINALITY THE NUMBER OF ITEMS IS IRRELEVANT – IT'S HOW THE SET MATCHES UP AGAINST ANOTHER.

Imaginary numbers

It may seem like playing with words to define an infinite set
as one that has the same cardinality as a subset, something
detached from reality, but we have to remember that mathematics is not about the real world. It's the logical following of a set
of arbitrary rules – the axioms on which a system of maths is
based. Even unreal mathematical concepts can be of value in
the real world, though. A good example is **imaginary numbers***.
These are based on the outcome of a seemingly simple question – what is the square root of a negative number? What, for
example, is $\sqrt{-1}$?

I HAVE AN
IMAGINARY
NUMBER OF
IMAGINARY
FRIENDS ...

We are looking for the number that, multiplied by itself, gives –1. But we know that both 1 x 1 and –1 x –1 are 1. Neither is the square root of –1. So mathematicians assign a value of i to the square root of –1. A whole structure of mathematics has been built on these imaginary numbers, and on **complex numbers***, combining real and imaginary, such as 3 + 2i.

Engineering and physics make heavy use of imaginary numbers, as long as the final result eliminates them, because they are an easy way to extend the number line into two dimensions. The real numbers form a traditional number line running horizontally and the imaginary numbers another number line running vertically. Any point on the two-dimensional space is then identified by a complex number.

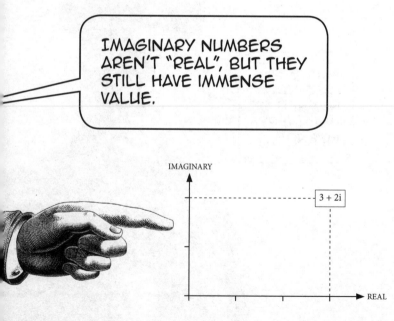

IMAGINARY NUMBERS AREN'T "REAL", BUT THEY STILL HAVE IMMENSE VALUE.

Aleph null

Once we're taking a set theory approach to infinity – true infinity,
rather than Aristotle's potential infinity – it needs a different symbol. Cantor chose aleph, the first letter of the Hebrew alphabet,
and specifically he called the infinity of the counting numbers
aleph zero or aleph null ($\aleph_0$). This suffix was a good example
of the mathematician's view of the world. Anyone else might
assume that infinity is infinity. But mathematicians are hyper-
sceptics. We can't assume that all infinities are the same; there
has to be a clear identification of what is meant by infinity. Aleph
null is the basic infinity, the infinity of the positive integers.

אָלֶף־בֵּית עִבְרִי

THE CARDINALITY
OF THE COUNTING
NUMBERS IS $\aleph_0$ –
BUT IS THAT TRUE
OF ALL INFINITE
SETS?

There has been some dispute as to why Cantor chose aleph as the symbol for "true" infinity, just as no one is entirely certain why Wallis devised the lemniscate. It has been suggested that, though Cantor's family was Christian, he had Jewish roots and was aware of the mystical tradition of the Kabbalah, where one of the representations of the infinite Ein Sof was the letter aleph. There's no doubt that Cantor was influenced by religious imagery, and he would certainly have been aware of the use of "the alpha and the omega" in Christianity. It may be he was simply bored with Greek symbols.

Aleph null has some strange qualities. We can add 1 to it and still end up with the same value. You can see why this happens if you imagine putting the series 1, 2, 3 in one-to-one correspondence with x, 1, 2, 3... So 1 in the first series corresponds with x in the second, 2 in the first series with 1 in the second, and so on. You can go through the whole lot matching them off, so they have the same cardinality. What's more, add aleph null to itself and you get aleph null. (Because 1, 2, 3... and 1a, 1b, 2a, 2b... have the same cardinality.) For that matter, you can multiply aleph null by itself – and still get aleph null. Yet really it hardly seems surprising that this happens, because infinity is infinity.

$$\aleph_0 + 1 = \aleph_0$$

$$\aleph_0 + \aleph_0 = \aleph_0$$

$$\aleph_0 \times \aleph_0 = \aleph_0$$

$$1 \rightarrow x$$
$$2 \rightarrow 1$$
$$3 \rightarrow 2$$
$$4 \rightarrow 3$$

and so on...

Cardinals and ordinals

When Cantor devised aleph null he was thinking of cardinal numbers, the numbers that define the size of a set. Yet the number 7, say, has more than one application, which would lead to a second kind of infinity. The symbol 7 may mean the cardinal value "7" as in "I have 7 oranges". But 7 is also an **ordinal***. "Cardinals" and "ordinals" sound vaguely religious, but ordinal just means a number dependent on *order*.

If I have a row of oranges, I can say: "This bit of the row contains 7 oranges" – a cardinal value. But I can also say: "This is orange number 7" – its ordinal value.

THIS ORANGE
HAS CARDINAL
VALUE 1, BUT
ORDINAL VALUE 7.

Although we're used to thinking of integers in order, the full set of integers can't have ordinal values. This is because to have ordinal values, every subset of a set has to have a first value. You might think "Surely every set has a first value?" But the integers don't. Think of the number line of all the integers.

IT LOOKS SOMETHING LIKE ... -5 -4 -3 -2 -1 0 1 2 3 4 5 ... BUT WHAT IS THE FIRST VALUE? WE CAN'T ASSIGN ONE.

-4 -3 -2 -1 0 1 2 3 4 5

On the other hand, the counting numbers, the positive integers, do have a first value, so they also have ordinal values.

Ordinal infinity

For finite numbers there's no obvious distinction between cardinals and ordinals. But they diverge at infinity. We know that $\aleph_0 + 1 = \aleph_0$ but this can't be true with the ordinal infinity – order continues. For ordinal infinity, Cantor resorted to traditional Greek symbols, using the "ultimate" omega, ω, as the limit of the ordinal list 0, 1, 2, 3, 4… This would then be followed by $\omega + 1$, $\omega + 2$… and so on. The mathematics of ω can be a little tricky. This is because $\omega + 2$ is not the same as $2 + \omega$, and $\omega \times 2$ is not the same as $2 \times \omega$.

The reasoning goes something like this. You can represent ω + 2 as {1, 2, 3… ω₁, ω₂} where ω_1 and ω_2 are the next two values after ω. And 2 + ω is {ω₁, ω₂, 1, 2, 3…}. In the latter case any initial segment of the set is smaller than an infinite set, so 2 + ω = ω. But for ω + 2 the segment before ω_1 is infinite, so the whole set is bigger than ω. The ωs build up through ω^ω and $(\omega^\omega)^\omega$ to ω raised to the power of ω for ω times. This is (arbitrarily) given the name ε_0, and so it continues.

ω IS JUST THE START.

Cantor developed a hierarchy of *ordinal* infinities, but was aleph null the *cardinal* limit?

111

Countably infinite

Cantor wanted to check how flexible aleph null really was. $\aleph_0$ is the cardinality of the counting numbers and the squares, or for that matter the odd or even positive integers. These sets are called **countably infinite*** or **denumerable**. At first sight this term "countably infinite" is an oxymoron. By definition, something infinite can't be counted. Apart from anything else, as we saw with the integers and the squares, such a set can be put in a one-to-one correspondence with a subset. How can you count anything like that? But countable just means having the same cardinality as the counting numbers.

1, 2, 3, 4... INFINITY!
AH AH AH AH AH AH!

Cantor's elegant proof

Was aleph null the cardinality of all infinite sets? Were they all countably infinite? Cantor started with rational fractions, fractions made out of the ratio of whole numbers. He was to prove that there were also aleph null of these, using a delightfully neat proof that requires no maths.

Imagine laying out every rational fraction in a huge table. We're actually repeating many of the fractions. If you look down the diagonal, they're all 1. It doesn't matter that we've got redundancy: the table, continued for ever in both directions, has every single ratio in it.

1/1	2/1	3/1	4/1	5/1	6/1	7/1	8/1	9/1	10/1	...
1/2	2/2	3/2	4/2	5/2	6/2	7/2	8/2	9/2	10/2	...
1/3	2/3	3/3	4/3	5/3	6/3	7/3	8/3	9/3	10/3	...
1/4	2/4	3/4	4/4	5/4	6/4	7/4	8/4	9/4	10/4	...
1/5	2/5	3/5	4/5	5/5	6/5	7/5	8/5	9/5	10/5	...
1/6	2/6	3/6	4/6	5/6	6/6	7/6	8/6	9/6	10/6	...
1/7	2/7	3/7	4/7	5/7	6/7	7/7	8/7	9/7	10/7	...
1/8	2/8	3/8	4/8	5/8	6/8	7/8	8/8	9/8	10/8	...
1/9	2/9	3/9	4/9	5/9	6/9	7/9	8/9	9/9	10/9	...
1/10	2/10	3/10	4/10	5/10	6/10	7/10	8/10	9/10	10/10	...
...	...	...	...	...	...	...	...	...	...	...

Next, Cantor set up a repeating path through the table. In this case, it's: "Move one to the right, go down diagonally left until you hit the edge, go one down, go diagonally up right until you hit the edge. Then repeat." Finally, he put each item in that path in one-to-one correspondence with a counting number. He had effectively proved that aleph null applies to the rational fractions as well – they have the same cardinality as the positive integers because they match one-to-one by going through this sequence.

THOUGH THIS IS THE PATH I USED, IT'S NOT THE ONLY ONE THAT EXISTS.

start here

THE POINT IS THAT THERE'S A MECHANISM TO SET UP THE ONE-TO-ONE CORRESPONDENCE.

Covering the number line

The infinity of rational fractions has a surprising quality. Imagine a number line from 0 to infinity, like a ruler, with every rational fraction marked on it. Our aim is to cover the whole number line. We issue each rational fraction with an umbrella (a simple T shape). The first umbrella is ½ a unit in width, the second umbrella is a ¼ unit, and so on. Each umbrella stretches to a rational fraction either side of the one holding it, so they cover the whole number line. But ½ + ¼ + 1/8… adds up to 1. So a set of umbrellas only 1 unit in width covers the infinite number line of rational fractions.

Another Cantor proof

When we speak of mathematical proofs, we think of pages of impenetrable equations. When British mathematician Andrew Wiles proved Fermat's Last Theorem in 1995, his proof ran to over 100 pages. Yet Cantor's proof of the cardinality of the rational fractions is so simple that it's easy to think of it as a trite truism that doesn't provide any insights. (To be fair, this is a simplified presentation – to be rigorous requires more than just "We can find a path and put it in one-to-one correspondence".)

IT WOULD SEEM OBVIOUS THAT ALL NUMBERS COULD BE TREATED SIMILARLY. BUT CANTOR COULDN'T LEAVE IT THERE.

As the Pythagorean Hipparsus found out to his cost, rational fractions aren't the only kind of non-whole numbers. There are also the irrationals, the numbers that written as decimals go on for ever and ever. Does the full set of these also squeeze into aleph null? With another blindingly simple proof, Cantor was to show that this *wasn't* the case. He imagined putting every single decimal, rational and irrational, between 0 and 1 into a list. If he could achieve that, then he could use exactly the same one-to-one proof, matching each decimal against its position in the list, and would prove that this was another aleph null set.

0

0.000000... erm

To work this proof, Cantor needed to be able to study sequential numbers in the list. If they are in order that's impossible, because the first number is 0.000... all the way to infinity with 1 at the end, the second is the same with 2 at the end, and so on. So Cantor scrambled the list and picked out the first few numbers.

> LET'S LOOK AT THE DIAGONAL THROUGH THOSE NUMBERS – THE FIRST DECIMAL PLACE OF THE FIRST NUMBER, THE SECOND OF THE SECOND AND SO ON.

0.21584032048303404035930...
0.92939212493373921239446...
0.52030202578403024842231...
0.65873032294825193482488...
0.15740302049304958102733...
0.33335939320293919290111....

...

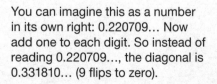

You can imagine this as a number in its own right: 0.220709... Now add one to each digit. So instead of reading 0.220709..., the diagonal is 0.331810... (9 flips to zero).

Finally, Cantor compared this new number 0.331810… with the original table. It's not the first number, because they differ in the first digit. It's not the second number, because they differ in the second digit. It's not the third number. And so on. He had generated a number that doesn't appear in the list. Cantor had proved with beautiful simplicity that you can't cram all the decimals between 0 and 1 into a list with cardinality aleph null. The count of these decimals was something bigger – something bigger than infinity.

0.331810…

WE'RE TALKING ABOUT **TRANSFINITE NUMBERS***. TAKE A MOMENT TO THINK ABOUT THAT.

0.220709…

Points in space

Here's one other remarkable thing Cantor discovered. Up to now we've been working on a number line, a one-dimensional list of numbers, and have come up with this new infinity – Cantor referred to it as the "infinity of the continuum" or $\aleph_c$, since it covered every number in the continuous spectrum between 0 and 1. It's every point on a line. But how many points are there in a square, or a cube? The last of Cantor's simple proofs extends aleph to the points in space. We'll work it just for a square, though you can apply this to any number of dimensions.

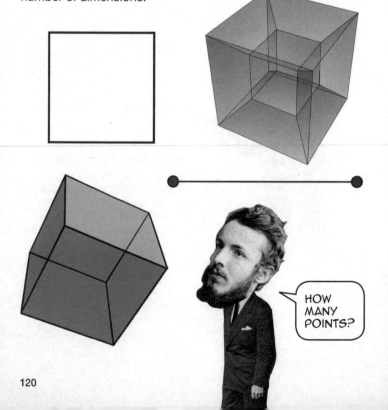

HOW
MANY
POINTS?

We define a position on a line with one number. So, for instance, on a number line running from 0 to 1, the half-way position is 0.5. Similarly, to define a point on a square we use two numbers. On a map, these are grid references or latitude and longitude; on a graph, X and Y coordinates. These are "Cartesian coordinates" after the French philosopher **René Descartes** (1596–1650) who demonstrated the way algebra and geometry come together in such a plot.

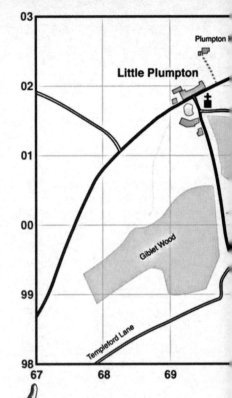

Using two numbers to locate a point in a plane was not Descartes' invention, though – **Ptolemy** (c. AD 90–168) had maps using two coordinates in his AD 150 world atlas.

121

In a square that, like our original number line, runs from 0 to 1 on both X and Y sides, we can refer to any point using two numbers between 0 and 1, the X coordinate and the Y coordinate. What Cantor spotted was that we don't need two numbers to refer to that point. We can create a new number by alternating the digits in the two values. So, for instance, a point identified as 0.5921 on the X axis and 0.2843 on the Y axis is uniquely identified by 0.**52982413**, where the odd decimal places identify the X axis location and the even the Y.

When they told you at school that you needed two numbers to identify a point on a two-dimensional plane, they were wrong. Every point is identified by a single value. Of course there's more information in the new number – it has twice as many decimal places – but it's still a single number. And so the cardinality of points in a square is the same as that of the continuum between 0 and 1, $\aleph_c$. The same argument applies to the points in a cube, or an n-dimensional hypercube.

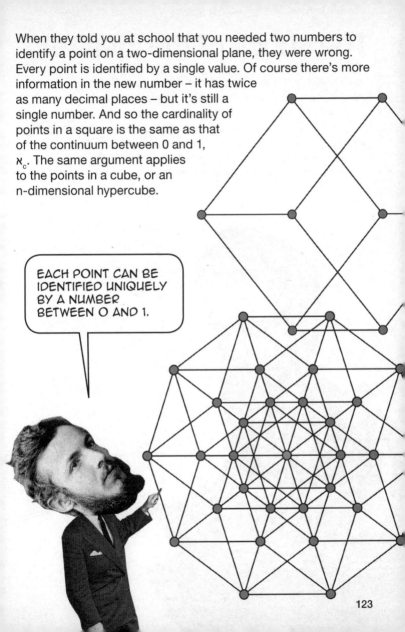

EACH POINT CAN BE IDENTIFIED UNIQUELY BY A NUMBER BETWEEN 0 AND 1.

The shock of the infinite

Cantor seems to have been more shocked by this discovery
– that $\aleph_c$ applied to any dimension of space – than his other
remarkable proofs. It's difficult to understand why this was so
shocking to him. Most minds are boggled by the concept of bigger and smaller infinities, while there's something more tangible
in the idea that the number of points on a line and the number of
points in a three-dimensional space (say) are identical.

I SEE IT BUT
I DO NOT
BELIEVE IT.

Some of the things Cantor had done certainly stretch the mind, but it wasn't anything we've seen so far that drove him to insanity. In part it was the frustration that one aspect of infinity would elude him for the rest of his life. He had shown that the infinity of the continuum, $\aleph_c$, all the fractions between 0 and 1, was bigger than aleph null, but was it $\aleph_1$, the count of aleph nulls, just as you could say that $\aleph_0$ is the count of the positive integers? It seemed reasonable – but Cantor couldn't find a proof for this, an idea that became known as the **continuum hypothesis**.

IS THE INFINITY OF THE CONTINUUM ALEPH ONE?

Power sets

Cantor did take one step in this direction, to understand which we need the concept of **power sets**. Let's take a very simple set of three items – knife, fork and spoon. It's a set of cardinality 3. But it has more subsets. We can identify:

nothing (the empty set)

knife

fork

spoon

knife + fork

spoon + fork

knife + spoon

knife + fork + spoon

A total of 8 subsets. It turns out that the cardinality of all the subsets of a set, called the power set, is always 2^c, where c is the cardinality of the set itself. In the case of our cutlery, that is 2^3, or 2 x 2 x 2, which is 8.

This feature of the power set even applies to the empty set, though that needs a little thinking through. This would make the cardinality of the empty set's only subset 2^0. But what is 2^0? We generally think of the power sign meaning "multiply by itself this number of times". So 2^3 is the same as $2 \times 2 \times 2$, or 8. But what is 2^0? It arises from the way power arithmetic works. If, for instance, you multiply 2^2 (2×2) by 2^3 ($2 \times 2 \times 2$) you get 2^5 ($2 \times 2 \times 2 \times 2 \times 2$). You add the powers. So $2^2 \times 2^0$ must be 2^2. Making 2^0 (or, for that matter, anything^0) 1.

THE CARDINALITY OF THE POWER SET OF THE EMPTY SET IS 1.

Cantor proved that the infinity of the continuum ($\aleph_c$) was the power set of the real numbers. He could not work out if $\aleph_c$ was $\aleph_1$, but he could show that it was $2^{\aleph_0}$, which he felt was a step in the right direction. You can see this by writing out all the numbers between 0 and 1 in binary. In binary 0.5, for example, is 0.1 followed by an infinite row of zeros, while 0.25 is 0.01 followed by an infinite row of zeros, and so on.

ANY VALUE BETWEEN 0 AND 1 CAN BE WRITTEN IN BINARY AS A ROW OF ALEPH NULL DIGITS THAT ARE EITHER 0 OR 1.

Whenever there's a set of things, each of which can have two values, the number of combinations is 2^n, where n is the number of things. So we have $2^{\aleph_0}$ possible numbers in the set of numbers between 0 and 1, which has the cardinality $\aleph_c$.

Cantor under attack

While he was under stress trying to confirm the continuum hypothesis, Cantor came under academic attack. A one-time mentor, **Leopold Kronecker** (1823–91), who was much more powerful in the academic establishment, set out to ruin Cantor because Kronecker simply could not stand the implications of Cantor's work. Kronecker was a purist. He was happy only with the existence of integers and numbers directly based on them, like rational fractions.

ANYTHING ELSE HE THINKS IS DUBIOUS – EVEN SOMETHING AS STRAIGHTFORWARD AS AN IRRATIONAL FRACTION.

AND CANTOR'S ALEPHS ARE SIMPLY BLASPHEMOUS!

Kronecker was determined to keep Cantor's work from academic acceptance.

For a while the battle between Cantor and Kronecker seemed as if it could go either way. When Cantor came up with the proof that the points in an n-dimensional space were of the same cardinality as the continuum between 0 and 1, he did manage to get it published in a leading German publication, *Crelle's Journal* – but only after many months of delay, which an editor at the journal confided was due to a barrage of negative comment from Kronecker.

Cantor tried to take the argument to Kronecker by applying for a professorship in Berlin, which he knew would enrage his Berlin-based rival.

KRONECKER WOULD FLARE UP AS IF STUNG BY A SCORPION, AND WITH HIS RESERVE TROOPS WOULD STRIKE UP SUCH A HOWL THAT BERLIN WOULD THINK IT HAD BEEN TRANSPORTED TO THE SANDY DESERTS OF AFRICA WITH ITS LIONS, TIGERS AND HYENAS.

As expected, Cantor's application to Berlin failed. Meanwhile, Kronecker dug a trap for his opponent. Cantor's papers were often published in the journal *Acta Mathematica*, run by a friend of Cantor's called Magnus Gösta Mittag-Leffler. Kronecker offered his own papers to Mittag-Leffler. This was a huge goad for Cantor – he had found a safe haven in *Acta Mathematica* and now his enemy was invading it. As Kronecker anticipated, Cantor played the prima donna, threatening to stop publishing with Mittag-Leffler if he considered Kronecker's paper.

THIS OF COURSE MADE RELATIONS BETWEEN CANTOR AND HIS ONLY PUBLISHING FRIEND STRAINED.

MEANWHILE, KRONECKER'S PAPER EVAPORATED AWAY.

Cantor succumbs

Under constant, renewed attack from Kronecker, trapped in the mathematical backwater of the university of Halle, unable to prove the so-called continuum hypothesis, Cantor's mind was shattered.

GEORG
FERDINAND
LUDWIG
PHILIPP
CANTOR
1845–1918

HE DIED IN 1918 IN A MENTAL CLINIC HE HAD REPEATEDLY HAD TO VISIT OVER THE PREVIOUS YEARS.

The irony is that Cantor's successors would prove that he was wasting his time in trying to pin down the relationship between $\aleph_0$ and $\aleph_c$. The first step on the route came from another man whose mental stability would be challenged by the contemplation of infinity, the German/Czech mathematician Kurt Gödel.

Gödel's shocking proof

Kurt Gödel (1906–78) devised the most shocking proof in mathematics. His masterpiece, the **incompleteness theorem**, states that in any system of mathematics there will be some problems that it's impossible to solve. A system is the series of axioms, or basic rules and assumptions, on which the maths is based. A crude approximation to Gödel's theorem is to imagine dealing with the statement: "This system of mathematics can't prove that this statement is true." Is this statement true?

> IF THE SYSTEM PROVES THE STATEMENT, THEN IT CAN'T PROVE IT. IF THE SYSTEM CAN'T PROVE THE STATEMENT, IT STILL CAN'T PROVE IT.

Whatever happens, this is an unprovable statement.

Like Cantor, Gödel was anything but a stable character. Though not a Jew, in the 1930s he found Nazi Austria an increasingly uncomfortable place to work. In 1939 he and his wife decided to escape to America. Unable to take the western route, they used the Trans-Siberian railway and reached San Francisco via Japan. But despite getting a position at the prestigious Institute for Advanced Study at Princeton, Gödel suffered from increasing paranoia. He was convinced that someone was trying to poison him and would eat only food prepared by his wife. When she died he effectively starved to death.

WHILE ON HOLIDAY, THE DISTRACTED GÖDEL WAS NEARLY ARRESTED AS A SPY AS HE PACED ALONG THE SEAFRONT, MUTTERING IN GERMAN TO HIMSELF.

THE LOCALS THOUGHT HE WAS WAITING TO CONTACT A GERMAN U-BOAT.

Back to the continuum hypothesis

Gödel managed to prove that the continuum hypothesis was not inconsistent with set theory, but his mental state became too unstable to ever apply his work further to Cantor's problem with infinity. It was another, younger mathematician, **Paul Cohen** (1934–2007), who showed that it would never be possible to either prove or disprove the continuum hypothesis. No one can be certain if $\aleph_c$, the infinity of the continuum from 0 to 1, is the same as $\aleph_1$.

INSTEAD, I WAS ABLE TO PROVE THAT THE CONTINUUM HYPOTHESIS IS INDEPENDENT OF THE AXIOMS OF SET THEORY – IT WORKS TOTALLY OUTSIDE THEIR BOUNDS.

The French-born mathematician **André Weil** (1906–98), who like many others moved to the USA during the lead-up to the Second World War, perhaps best summed up the frustration generated by the kind of result we're left with after Gödel and Cohen's work on the continuum hypothesis. It is not inconsistent with set theory, and yet it can never be linked with set theory. It can neither be proved nor disproved. As long as we stick with the same axioms used as the foundation for set theory, it will never be possible to make any further progress.

GOD EXISTS SINCE MATHEMATICS IS CONSISTENT, AND THE DEVIL EXISTS SINCE WE CANNOT PROVE IT.

Does infinity exist?

Few mathematicians were as fussy as Kronecker, but he wasn't the only one to be uncomfortable with Cantor's revelations on infinity. A contemporary once remarked that Cantor's ideas "appear repugnant to the common sense". In the end we're always plagued with the uncertainty: does infinity truly exist, or is it merely a convenient – as Aristotle would have it, a *potential* – concept?

IT'S HARD TO SAY IF THERE'S A TRUE INFINITY IN THE REAL WORLD.

LIKE ARISTOTLE, WE HAVE TO ASK: DOES TIME STOP OR END? IS THERE A POINT WHERE WE CAN NO LONGER DIVIDE TIME OR SPACE?

Fractal infinity

One application that hints at a real infinity is **fractals**. These provide a way to produce an infinitely long path in a finite space. The simplest approach is a Koch curve, described by Swedish mathematician **Helge von Koch** (1879–1924) in 1906.

It starts with an equilateral triangle. We put a 1/3-sized triangle in the middle of each face of the triangle, pointing outwards. We now have a longer circumference. Now put new triangles 1/3 the size of those extra triangles on each face. And so on. The shape, sometimes called a Koch snowflake, has a circumference that heads towards infinite length but never emerges from a circle that encloses the original triangle.

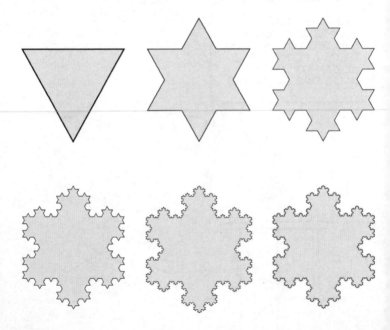

When you zoom in to the detail of a fractal, it resembles the larger whole. This "self-similarity" is one of the defining characteristics of the form. Fractals emerged from the work of Cantor and others, but did not get the name until 1975 when Polish/French/American mathematician **Benoît Mandelbrot** (1924–2010) devised the term to emphasize that the results of these mathematical functions were "fractured". Mandelbrot is probably best known for the **Mandelbrot set**, a particular fractal form that became a poster image of the 1980s.

THE BOUNDARY OF THE SET THAT PRODUCES THIS STRIKING IMAGE GETS MORE AND MORE COMPLEX AS MORE DETAIL IS ADDED.

Recursion

A fractal is generated by repeatedly applying what is usually a (relatively) simple equation to build a structure that becomes more and more complex. It's no coincidence that their popularity arose in the 1970s and 80s as computers were becoming standard tools in mathematics. The repeated application of a formula – recursion – is a natural mechanism for computing and while the initial stages of the Cantor set (see page 145) and the Koch curve can be produced by hand, a structure like the Mandelbrot set required computers to reach a meaningful result.

IT WAS ALSO REALIZED THAT MANY NATURAL FORMATIONS ARE LIKE FRACTALS.

Fractals in nature

Although fractals can seem to be abstract mathematical forms – attractive but useless – they actually mirror natural formations. Trees, mountain ranges, snowflakes, clouds are all roughly fractal. It's not that there's some inherent fractal aspect to nature, but rather that the way these natural objects are formed involves a similar repeated simple process. The natural objects are only roughly fractal because they don't follow rigid mathematics – there are many influences that can change the outcome – but they usually have a degree of the self-similar property common to many fractals.

SOME OF THESE NATURAL FORMS – FERNS, FOR EXAMPLE, AND MOUNTAIN RANGES – ARE EASY TO SIMULATE WITH SIMPLE FRACTAL FORMULATIONS.

When fractals first became popular it was felt that they were bound to have many applications. At the time, photographs were becoming common on computers, and they took up a lot of the limited disk space available. Fractal researchers set up a company, Iterated Systems, to sell compression software for images.

IN PRINCIPLE, FRACTAL COMPRESSION WAS BETTER THAN THE DOMINANT JPEG FORMAT, BECAUSE THE IMAGES DIDN'T BECOME BLOCKY ON ZOOMING IN, JUST INCREASINGLY BLURRY.

But the technology never really caught on and has remained niche. Fractals do have a number of other uses, both analytical and functional, but are yet to really break out.

Measuring the coastline

The Koch curve and the Mandelbrot set come close to the real-world problem of defining the length of the coastline of an island like Britain. If you imagine measuring around the coastline using a metre rule, you would come up with one figure.

BUT USE A SMALLER RULER THAT CAN GO MORE INTO THE CRACKS AND CRANNIES, AND YOU'LL GET A LONGER DISTANCE.

Until we reach atomic limits, the length can get as long as you like. We're used to science being able to come up with precise values, but here's a measurement that doesn't have a specific value. The answer can only ever be: "It depends."

In practice, when measuring the coastline we always hit physical limits. Even with a measuring device that could distinguish infinitely small differences in size, the atoms making up the coast have a finite size that could be considered the limit of measurement. If we go beyond this, it appears that there's a distance, the Planck length, below which it is inherently impossible to measure. The Planck length is around 1.6×10^{-35} metres. It also implies a minimal unit of time, the time light takes to cover such a distance, about 5.4×10^{-44} seconds.

Max Planck (1858–1947)

SO MAYBE ARISTOTLE WAS WRONG ABOUT OUR ABILITY TO DIVIDE TIME AND SPACE FOR EVER.

The Cantor set

Cantor came up with a class of sets that bridges the infinity of the continuum ($\aleph_c$) and fractals. The best-known form, the Cantor ternary set, is formed by a straightforward repeated action on the continuum between 0 and 1.

Imagine the whole 0 to 1 number line as a solid line. Now chop out the middle third of that line, so you have two blocks, each 1/3 in length, either side of the gap. Next chop out the middle third of each of those blocks. So now you have four blocks. And so on for ever.

THE RESULT IS A PATTERN THAT GRADUALLY THINS OUT BUT NEVER DISAPPEARS.

It might seem that when the process is taken all the way to infinity, the pattern will disappear as it will be entirely eaten away, but there's a subtlety in the definition of the chunks that are taken out that prevents this from happening.

WHAT IS DELETED IS THE "OPEN" MIDDLE THIRD. THIS MEANS THAT IT DOESN'T INCLUDE THE END POINTS.

So when, for example, the first chunk is taken out from 1/3 to 2/3, both the points 1/3 and 2/3 are still there – the gap is infinitesimally smaller than the pieces left behind, providing enough material to keep the pattern going.

The Cantor set is a fractal. If you examine it at any level, it's similar to any other level, just like the Koch snowflake. This is a set that's a bit like the series $1 - 1 + 1 - 1 + 1 - 1 + 1...$ On the one hand, when it reaches infinity there seems to be nothing left. It appears that it will be eaten away to nothing, and it can be proved that there are no non-zero intervals – chunks that are more than a single mathematical point. However, there are an infinite set of these points – it has the same cardinality as the points in the original line, $\aleph_c$.

IT'S THERE AND IT'S NOT THERE ...

An infinite universe?

Fractals apart, is there physical infinity? For example, is the universe infinite? Throughout scientific history, the answer has alternated between yes and no. Ever since the ancient Greeks there have been arguments about the rights and wrongs of an infinite universe. Over time our understanding of the scale of the universe has grown and grown.

Nicolaus Copernicus
(1473–1543)

I CALCULATE THE UNIVERSE TO BE AROUND 90 TIMES BIGGER THAN THE SCALE ASSUMED BY ARCHIMEDES – ABOUT 0.001 LIGHT YEARS ACROSS.

By the 20th century, it was discovered that even our nearest major galactic neighbour, the Andromeda galaxy, was 2 million light years away.

We now know that the universe has to be at least 90 billion light years across. Although we can only see light that has been travelling for the lifetime of the universe, around 13.7 billion years, because the universe is expanding we can see about 45 billion light years in each direction. But we can't know for certain if it's finite.

> I BELIEVE THE UNIVERSE MUST BE INFINITE, BECAUSE IF IT WERE FINITE, OBJECTS NEAR THE EDGE WOULD FEEL GREATER GRAVITATIONAL ATTRACTION TOWARDS THE MIDDLE, SO THE WHOLE THING WOULD COLLAPSE.

Newton

For that matter, a finite universe could be a part of an infinite *multiverse*.

Edge of the universe?

As far back as Roman times it was argued that the universe had to be infinite, because otherwise there would be an edge, and what was outside? What would happen to an arrow shot through the edge? More recently it was realized that a finite universe doesn't have to have edges.

THINK OF A SPHERE. THE SURFACE IS FINITE, BUT WITHOUT EDGES.

It's possible to translate the same image into an extra dimension to produce a finite universe with no boundaries. But even if it did have an edge, we now think that nothing could cross the boundary, as the universe is expanding so quickly that nothing could catch it up.

One of the most pragmatic approaches to the infinity of the universe is supposed to have originated with Alexander the Great. One of the companions of the young Macedonian king on his expeditions into Asia was the philosopher **Anaxarchus** (c. 380–320 BC), a student of Democritus, the originator of the first atomic theory. Anaxarchus informed Alexander that there were an infinite number of worlds, and Alexander is said to have burst into tears. His friend asked him what was wrong and Alexander came back with a striking reply.

DO YOU NOT THINK IT A MATTER WORTHY OF LAMENTATION THAT WHEN THERE IS SUCH A VAST MULTITUDE OF THEM, WE HAVE NOT YET CONQUERED ONE?

Perhaps this is where modern generals go wrong: not enough of them employ philosophers on the front line.

As it is, we just don't know about the size of the universe. It might seem that the big bang theory makes the universe necessarily finite, as the whole thing is thought to have emerged from a point. But the big bang doesn't exclude our known universe from being an expanding bubble in a far larger, and quite possibly infinite multi-verse. And, for that matter, the big bang is just one of a number of theories that match the observed data, some of which do allow for a single infinite universe. We have no definitive proof either way.

Quantum infinity

Perhaps our best chance of seeing infinity for real may emerge from the nascent science of quantum computing. This is computing that relies on a bit that is not a simple 0/1 switch, as in a conventional computer, but rather the state of a quantum particle like a photon of light or an electron. Quantum particles obey a whole different set of rules to macro-sized objects like people.

> SPECIFICALLY, A QUANTUM PARTICLE CAN BE IN MORE THAN ONE STATE AT A TIME.

Erwin Schrödinger (1887–1961)

This is probably most easily demonstrated with an experiment that dates back to the 19th century, called Young's slits.

The slit experiment

In this experiment, devised by **Thomas Young** (1773–1829), light is sent through a pair of slits. The two beams of light meet behind the slits and interact. If you think of light as **waves**, this "interference" is like two sets of waves crossing each other. Some bits of the waves will reinforce each other – both going up at the same time, producing a particularly strong peak. Some will counteract each other – going in opposite directions – leaving a calm patch. Similarly, the light from the two slits interferes to produce a set of light and dark bands on a screen behind them.

I USED THIS EXPERIMENT TO DISPROVE THE IDEA THAT LIGHT IS MADE UP OF *PARTICLES*.

However, in the 20th century it was discovered that light really is made up of particles – *photons*. We can send photons through Young's slits one at a time. And the result is exactly the same – bands of light and darkness build up on the screen at the back. This can work only if photons go through *both* slits and interfere with themselves.

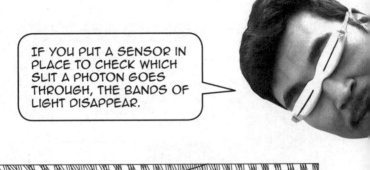

IF YOU PUT A SENSOR IN PLACE TO CHECK WHICH SLIT A PHOTON GOES THROUGH, THE BANDS OF LIGHT DISAPPEAR.

So these quantum particles are in *multiple locations*, and it's only the act of observing them that forces them to go through either one slit or the other.

Spin

For a quantum computer it's easiest to use another multiple quantum peculiarity, **spin**. Quantum spin isn't really about a particle spinning around: it's just an analogy. Spin is a property of all quantum particles. When it's measured, spin comes out as one of two values: "up" or "down".

> BEFORE MEASUREMENT, WE CAN'T SAY WHAT THE SPIN IS – WE JUST GIVE IT PROBABILITIES. IT MIGHT BE 47% UP AND 53% DOWN.

> UNTIL THE MEASUREMENT IS TAKEN, THE PARTICLE IS IN BOTH STATES WITH THE RELEVANT PROBABILITIES.

You can think of the ratio of probabilities as a **direction**. If the probabilities are 50:50 then the direction is half-way between up and down, but for any other values it's a different direction.

50

50

47

53

If a quantum computer can make use of this spin state as a quantum bit – a *qubit* – then we have a computer that isn't just handling 0 or 1, but could deal with an infinitely long decimal, the number that specifies the exact direction between the two spin states.

THIS ISN'T EASY TO USE, THOUGH. WHENEVER YOU MEASURE THE VALUE OF A QUBIT YOU DESTROY ITS STATE (AS WITH THE PHOTONS AND THE SLITS), AND THEY CAN ALSO COLLAPSE THROUGH INTERACTION WITH OTHER PARTICLES AROUND THEM.

But in principle a quantum computer is a real device that handles true infinitely long decimals, rather than the approximation used in a traditional computer.

The infinitesimal

When thinking of the infinite, we're never far from its inverse, the infinitely small. Although at opposite ends of the mathematical spectrum, infinity and the infinitesimal are inevitably bound together. The infinitesimal may not generate the same sense of awe – infinity produces more of a sense of wonder – yet the one is simply the inverse of the other.

AND AS WE'VE SEEN WITH FLUXIONS AND CALCULUS, WHILE BRUTE INFINITY MAY HAVE THE EXCITING ALEPHS AND OMEGAS, INFINITESIMALS ARE THE ASPECTS OF INFINITY THAT ARE MOST LIKELY TO BE PUT TO USE IN EVERYDAY LIFE.

Calculus gets around Bishop Berkeley's concerns about "the ghosts of departed quantities" by using indefinitely small values, rather than the true inverse of infinity, but in the 1960s, Israeli mathematician Abraham Robinson showed that, just as imaginary numbers could be of value even though technically they don't exist, so infinitesimals could be useful mathematical tools provided everything is tidied up at the end.

THERE'S NO MORE NEED TO TRY TO WORK OUT HOW TO FIT INFINITESIMALS INTO THE NORMAL SCHEME OF THINGS THAN THERE IS TO PUT $\sqrt{-1}$ ON AN ORDINARY NUMBER LINE.

Instead infinitesimals could be treated separately with their own mathematical operations.

Non-standard analysis

It wasn't, of course, just a matter of saying, "Let's call infinitesimals a different kind of number". Robinson used a technique called model theory to show that the same approach that gives a formal structure of arithmetic operations on real numbers can be stretched to include the infinitely large and the infinitely small. Robinson's approach became known as **non-standard analysis**.

> IT GIVES A MEANS TO ACCEPT INTUITIVELY OBVIOUS POSSIBILITIES – LIKE NEWTON'S FLUXIONS – BUT GIVING THEM THE RIGOROUS TREATMENT THAT MATHEMATICIANS DEMAND.

No longer was it necessary to worry about dividing by zero as the infinitesimal faded to nothing – you were dealing with acceptable if non-standard mathematical quantities.

In non-standard analysis, infinitesimals sit on a number line (mathematicians call the special number line including infinites and infinitesimals the *hyperreal number line*) and are bigger than −*a* but smaller than *a* for all values of *a*. They hover between the smallest negative and the smallest positive. Zero is the only real infinitesimal, but non-standard analysis brings in a whole cloud of other, non-real numbers sitting in the gap between −*a* and *a*. Most of us taught calculus will be surprised by the validity of infinitesimals. Non-standard analysis is something that many non-mathematicians still don't realize exists. But these are well-established mathematical techniques.

THINK OF A NUMBER, ANY NUMBER. IT'S SMALLER THAN THAT.

Infinitesimals and Brownian motion

An example of the application of non-standard analysis is in the modelling of Brownian motion. Back in the 1820s, botanist **Robert Brown** (1773–1858) noticed that pollen grains danced around when viewed in a drop of water under the microscope.

> INITIALLY I THOUGHT THIS WAS THE RESULT OF SOME KIND OF LIFE FORCE, BUT I FOUND THE SAME THING HAPPENED WITH TINY PARTICLES OF MATTER THAT WERE DEFINITELY DEAD.

By the 1870s, this movement was correctly explained as being due to impacts from randomly jiggling water molecules, and in 1905, Einstein developed a mathematical model of what was happening – but this wasn't enough.

Einstein's model of Brownian motion (one of his three landmark papers from 1905 that also included the first outing on special relativity and his ideas on the photoelectric effect that led to quantum theory) was a *statistical* model.

IT WAS GOOD AT PREDICTING THE OVERALL EFFECT, BUT IT COULDN'T FOLLOW WHAT HAPPENED TO INDIVIDUAL MOLECULES.

In the 1970s, American mathematician Robert Anderson used non-standard analysis to map out infinitesimally small movements that proved to be the only way to produce a workable model of Brownian motion at the detailed level. Those non-existent quantities were beginning to pull their weight.

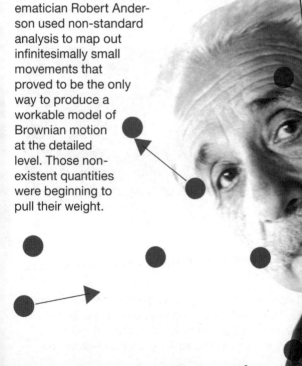

Robinson's work concentrated on infinitesimals, but it does also bring infinity into the hyperreal number line. It might seem at first that Cantor's transfinite numbers were coming out of the closet and joining the rest of maths, but this is not the case. Cantor's alephs won't sit on a number line like Robinson's infinities. The two systems are incompatible.

IT'S A BIT LIKE LOOKING AT A PHOTOGRAPH OF A 3D OBJECT. IMAGINE LOOKING AT PICTURES OF A SNAKE TAKEN FROM FRONT AND SIDE. THE PHOTOGRAPHS WOULD APPEAR TO BE DIFFERENT THINGS.

The same appears true of infinity. Look from one direction and you get alephs, from another, non-standard analysis. Some mathematicians believe that, far from conflicting, these two approaches will yield important results from their interaction.

Hilbert's Hotel

What certainly is real, however is the fascination of this subject. Take two classic paradoxes of infinity. The first is Hilbert's Hotel, named after the German mathematician David Hilbert. Hilbert's Hotel has one unique feature. It has aleph null rooms – one for each of the infinite set of counting numbers. Now imagine you arrive late one day. It's the only hotel in town. "Sorry", says the desk clerk, "we're full." "No problem", you say.

MOVE THE PERSON IN ROOM 1 TO ROOM 2.

MOVE THE PERSON IN ROOM 2 TO ROOM 3.

And so on right through the hotel. Now everyone has a room, but room 1 is free for you to occupy.

It looks like everything's fine, but then a special coach turns up. It's a coach with an infinite set of passengers on board. Again, the desk clerk has to apologize. The hotel is entirely full. Luckily, you're still at reception and able to take charge. "No problem", you say.

MOVE THE PERSON IN ROOM 1 TO ROOM 2.

MOVE THE PERSON IN ROOM 2 TO ROOM 4.

MOVE ROOM 3 TO ROOM 6.

And so on, doubling up room numbers throughout the hotel. Now all the odd-numbered rooms – an infinite set of them – are free for the new guests. Excellent.

Boring people may argue that Hilbert's Hotel could never exist. There are only a finite number of atoms in the universe (at least the basic big bang universe), loosely estimated at around 10^{80}. Once you have used up all those atoms you can't build any more rooms, so the hotel has to be finite.

FOR THAT MATTER, IT WOULD TAKE INFINITE TIME TO SHIFT THE GUESTS AROUND TO ACCOMMODATE THE NEW ARRIVALS – WHICH ISN'T ENTIRELY PRACTICAL.

But that's not the point. Even so, Hilbert's Hotel is somewhat predictable once you're familiar with the basics of transfinite arithmetic – but not all paradoxes of infinity are so tractable.

Gabriel's Horn

The second paradox, even more delightful to contemplate, is Gabriel's Horn. This is the mathematical structure produced by plotting a graph of 1/x (plotting values of 1/x, so when x is 1, y is 1, when x is 2, y is 1/2, when x is 3, y is 1/3 and so on) for every value of x greater than 1, and then spinning the resultant curve around the axis.

Imagine taking this curve as a sheet of paper and rotating it through 360 degrees around the vertical axis … or using the curve as a template on a lathe to cut out a three-dimensional object. The result is a shape like a straight hunting horn, but the pointy bit heads off to infinity.

Now the volume of Gabriel's horn can be calculated – it's pi: 3.14159 and so forth. If you're wondering how something can have a volume of π, remember the shape is for every x greater than 1. If that's 1 metre, the volume is pi cubic metres – 1 mile it's pi cubic miles and so on.

It's quite fun that this infinitely long horn has a finite volume, but that's reminiscent of the series 1 + ½ + ¼… adding up to 2. What sends a shiver down the spine is that Gabriel's Horn's surface area is infinite. If we hold it point down, we know exactly how much paint it would take to fill Gabriel's horn. Pi units.

> BUT HOWEVER MUCH PAINT WE HAVE, WE CAN NEVER COVER THE HORN WITH IT, BECAUSE IT HAS AN INFINITE SURFACE AREA. SPOOKY.

Mathematicians say that this isn't an issue, because volume and area are two different things – but the joy is in infinity's ability to tease us.

Just like Hilbert's Hotel, any attempt at a physical explanation of Gabriel's Horn runs into technical problems. The horn gets narrower and narrower. Before long (at least in terms of an infinitely long horn) it will be so narrow that its diameter is less than the size of a molecule of paint.

> AT A CERTAIN DISTANCE DOWN THE HORN YOU SIMPLY WOULDN'T BE ABLE TO FIT ANY PAINT ON THE SURFACE, SO YOU WOULD HAVE TO STOP PAINTING HAVING USED ONLY A FINITE AMOUNT OF PAINT.

But it still remains mindboggling that a surface that contains just pi units within it is infinite in area.

The jungle of infinity

Infinity is like a wild animal, spotted in the depths of a forest. You catch a glimpse of something, but moments later you aren't sure if you saw it at all. Then, unexpectedly, the animal comes into full view. A real problem with infinity is the dense undergrowth of symbols and jargon that mathematicians throw up. But the jargon is there for a good reason.

IT'S NOT PRACTICAL TO HANDLE THE SUBJECT IN DETAIL WITHOUT THESE NEAR-MAGICAL INCANTATIONS.

But the aim of this book has been to open up clear views on this most remarkable of mathematical creatures. Please continue to enjoy the world of infinity.

Glossary

Algorithm – a set of rules for solving a mathematical problem.

Calculus – mathematical technique based on infinity to deal with the way one value changes with another, or to work out areas and volumes.

Cardinal numbers – the counting numbers, identifying how many of an item there are in a set.

Cardinality – the property of a set that defines how big it is. Two sets have the same cardinality if the items in the set can be put in a one-to-one correspondence.

Complex numbers combine a real number and an *imaginary number*, such as 2+3i. Can be represented as a point on a two-dimensional graph, with real numbers on one axis and imaginary on the other.

Converging series – one where the total of the series is finite.

Countably infinite – a set that can be put in one-to-one correspondence with the counting numbers. Also known as *denumerable*.

Differentiation – using *calculus* to work out the way one variable changes with another. If the second variable is time, it's working out the rate of change.

Diverging series – one where the total of the series is infinite.

Empty set – set containing no items; corresponds to zero.

Fluxions (method of) – Newton's formulation of *calculus*.

Imaginary numbers – numbers based on the square root of –1. This has no real value, but is given the value i.

Indivisible – something that has been divided repeatedly and can no longer be divided. The numerical equivalent of an atom.

Integer – a whole number (no fractions) that can have negative or positive values.

Integration – using *calculus* to work out the area under a curve or the volume of a three-dimensional object. The reverse of *differentiation*.

Irrational number – one that can't be made from a ratio of two whole numbers, such as the square root of 2.

Natural number – a positive whole number. Natural in the sense that you can have this number of objects.

Number line – a sequence of numbers (like the *integers*) running along a line, like the markings on a ruler. Conventionally runs horizontally.

Ordinal numbers – the ordering numbers. Defines the position of an item in a sequence.

Series – a sequence of numbers added together.

Set – a collection of items. Set theory builds on the properties of sets to generate the rules of arithmetic.

Subset – a set that is a part of another set; e.g. odd numbers are a subset of the *integers*.

Tally – means of keeping track of counting using repeated marks, derived from counting on fingers.

Transcendental number – an *irrational number* that can't be calculated using a finite equation.

Transfinite number – a number that goes beyond aleph null, the infinity of the *integers*.

Unknowable number – a number that can't be calculated in any way short of writing it out digit by digit.

Further reading

A Brief History of Infinity – Brian Clegg – covering significantly more ground in an entertaining tour of infinity's role throughout history.

The Calculus Diaries - Jennifer Ouellette – a personal tour of the applications of calculus.

From Here to Infinity – Ian Stewart – mostly not about infinity, but an excellent tour of the heart of modern maths.

Isaac Newton – James Gleick – probably the best biography of this key figure in the development of calculus.

The Infinite Book – John D. Barrow – interesting book on infinity, particularly good on cosmology and applications, less strong on the maths.

The Mystery of the Aleph – Amir Aczel – good biography of Georg Cantor and summary of his work.

Understanding the Infinite – Shaughan Lavine – an academic title, but gives excellent background on the interpretation and understanding of infinity through history.

Zero – Charles Seife – engaging summary of the background and importance of infinity's arch rival.

Author's acknowledgements
Thanks to Duncan Heath for all the patient hand-holding and guidance, Oliver Pugh for the amazing illustrations, and Simon Flynn for introducing me to the series. A particular thank you to my inspirational maths teacher, Neil Sheldon, who got me interested in infinity in the first place.

Artist's acknowledgements
Thanks to Duncan Heath for getting me on board and effortlessly orchestrating this enormously enjoyable project. Thanks also to Brian Clegg (and Duncan again) for making my job a whole heap easier by providing a brilliant text to work with.

Brian Clegg is an award-winning popular science writer whose books include *A Brief History of Infinity*, *The God Effect*, *Before the Big Bang*, *Inflight Science*, *How to Build a Time Machine* and *The Universe Inside You*. He is a Fellow of the Royal Society of Arts and edits the www.popularscience.co.uk website.

Oliver Pugh is an award-winning graphic designer, illustrator and artist. Should the situation require it, he will design, draw or paint his way out.

Index